ADVANCES IN EDUCATIONAL PRODUCTIVITY

Volume 6 • 1996

OPTIMIZING EDUCATION RESOURCES

ADVANCES IN EDUCATIONAL PRODUCTIVITY

OPTIMIZING EDUCATION RESOURCES

Series Editor: HERBERT J. WALBERG
College of Education
University of Illinois at Chicago

Volume Editors: BRUCE S. COOPER
Graduate School of Education
Fordham University

SHEREE T. SPEAKMAN
K-12 Education Team
Coopers & Lybrand L.L.P.

VOLUME 6 • 1996

Emerald

United Kingdom – North America – Japan
India – Malaysia – China

Emerald Group Publishing Limited
Howard House, Wagon Lane, Bingley BD16 1WA, UK

British Library Cataloguing in Publication Data
A catalogue record for this book is available from the British Library

ISBN: 978-0-76-230082-2

CONTENTS

FOREWORD

Herbert J. Walberg

A recent report of the Organization for Economic Cooperation and Development (OECD, 1995) makes clear the need for this series, *Advances in Educational Productivity*, and particularly this book, *Optimizing Education Resources*. Among the rich nations for which data are available, the United States was highest in per-pupil schooling expenditures (adjusted for purchasing power). Yet American students from ages 9 to 14 made the least gains in reading proficiency. Clearly, these data and many other findings suggest that we can make substantial improvement in the efficiency of education finance and spending allocation.

With a highly original concept, Bruce S. Cooper and Sheree T. Speakman have attracted a distinguished group of scholars and financial analysts at the leading edge of education finance and management to contribute chapters to this book. Their data, means of analysis, and policy insights provide a rich source for future research and educational improvements. Going beyond mere juxtaposition, the editors show how chief ideas represented in the chapters relate to one another and have developed over the last few decades. The chapter authors themselves exhibit state-of-the-art analyses of great variety and power.

The aim of this series, however, is not to attain a consensus of ideas, values, perspectives, or methods. Reasonable analysts may and do differ on how much weight should be given to (a) the amount of money allocated for education, (b) equity of spending with respect to jurisdictions such as districts, schools, and classroom, (c) the mix of federal, state, and local contributions, and (d) the impact of funding on schools and pupils. Importantly, since the OECD data and several scholars have raised questions about the effects of spending on educational achievement, analysts have concerned themselves with how money should be spent and on what—rather than merely how much.

Thus, as the editors and I intended, a variety of views and research methods are represented in this (and other volumes in the series). The editors and chapter authors may not agree with one another on any given point. What we all agree on, however, is the importance of continuing research to improve the productivity of education services.

REFERENCE

Organization for Economic Cooperation and Development. (1995). Education at a glance. OECD indicators. Paris: OECD.

Chapter 1

INTRODUCTION:

TWENTY-FIVE YEARS LATER, *SERRANO* GOES TO SCHOOL

Bruce S. Cooper and Sheree T. Speakman

I remember that day 25 years ago: the halls buzzed with excitement in the Graduate Department of Education at the University of Chicago where I was a second-year doctoral student. "*Serrano, Serrano, Serrano v. Priest*!" was heard everywhere. The California Supreme Court had remarkably ruled that the unequal funding of students' education—based on the highly diverse property wealth of the communities—was unconstitutional: a violation of students' 14th Amendment right to "equal protection under the law."

The legal argument in the case, among many, had been developed by another University of Chicago graduate student, Arthur M. Wise, who was working under the same dissertation advisor as I, Professor Donald A. Erickson.

Advances in Educational Productivity, Volume 6, pages 1-14.

ISBN: 0-7623-0082-5

The legal rationale for the *Serrano* case (1971) had been spelled out in Wise's first book, *Rich Schools, Poor Schools* (1967). He had made several arguments for equity in addition to the "equal protection" tack that the California high court had used.

Oh, the optimism. We were sure that a new era of equal educational opportunity for those children trapped in poor communities was about to begin. Since the state court had turned to the Bill of Rights of the U.S. Constitution for justification for declaring the entire funding structure of California's public schools to be illegal, surely the same principle could be applied to interdistrict inequities in all other states.

Paul D. Carrington captured the mood of 1971, as he explained:

> Although the school-finance systems of every state but Hawaii were subjected to fundamental challenge, the decision was widely hailed on all sides. The press, legislative groups, educators at all levels in the administrative hierarchy, and taxpayer organizations were all enthused. Liberal civil rights adherents rejoiced at the apparent triumph of egalitarianism, and conservative property owners rejoiced at the apparent demise of the local property tax (Carrington, 1973, p. 162).

We were on the verge, we believed, of a revolution in education equality that would rival the impact of racial desegregation under *Brown* (1954, 1955). After all, if the very foundation of school finance in the United States—tax on local property—were unconstitutional, then the entire edifice would come crashing down, ushering in a new era, we believed, of true equality for all children—rich and poor, urban and suburban, white and of color.

EQUITY: A STATE FOCUS

Twenty-five years have passed since the original *Serrano* decision; and its children and grandchildren are all around us. But not in quite the way we had predicted on that heady day in August 1971 at the University of Chicago. First, the dream of *Serrano*'s bringing all the nation's schools under a 14th Amendment principle of "equal protection under the law" was decimated by the U.S. Supreme Court itself in the only major ruling of the high court in public school finance.

The high court confirmed in *Rodriguez* v. *San Antonio Independent School District*, 411 U.S. 1, 93 S. Ct. 1278 (1973) that, yes, money was in fact being spent unequally on K-12 education across districts in the state of Texas; that children living in property-poor districts were indeed being denied programs of equal quality to students in richer districts; and that public schools had the responsibility to educate all children to their full ability under the state's laws and authority. Justice Powell's statement captures the feeling of the court:

> We hardly need add that this Court's action today is not to be viewed as placing its judicial imprimatur on the status quo. The need is apparent for reform in tax systems which may have relied too long and too heavily on the local property tax. And certainly innovative new thinking as to public education, its methods and its funding is necessary to assure both a ***higher level of quality and greater uniformity of opportunity***. These matters merit the continued attention of scholars who already have contributed much by their challenges. But the ultimate solution must come from the lawmakers and from the democratic pressures of those who elected them (*San Antonio Independent School District* v. *Rodriguez*, 411 U.S. 1, 93 S. Ct. 1278 (1973), emphases added).

However, in its interpretation of this Texas case, *Rodriguez* (1973), the U.S. Supreme Court ruled that children living in a property-poor district were not a "suspect category" like race, sex, age, or religion, and thus were not protected by the Bill of Rights. After all, the San Antonio Unified School District had rich kids, poor kids, white kids and black and Latino children, girls and boys. Missing was the *category* of person being denied the protection of the U.S. Constitution. The case was referred to the state courts, where it has remained ever since, with each case being fought over the language in the particular state's constitution and laws.

Second, concerns for equity have since been joined with issues like school efficiency, productivity, and accountability. As Justice Powell specified in *Rodriguez*, new thinking is necessary to ensure both high quality and uniformity of opportunity. While education equity remains a key concern, so too are measures of how money is distributed, used, and the results of those expenditures for children's learning.

Thus, *Rodriguez* delayed the movement toward financial equity for two or more years, until an additional 29 states had their *Serrano* cases prosecuted under their own states' constitution (guaranteeing equal education for all—or not). Twenty states had court decisions: that their heavy reliance on property taxes to support public education led to an unconstitutional system of financial inequality—forcing these states to make greater effort to fund the poorer school districts—even to a point of penalizing the richer ones.

Third, *Serrano* and its successors have had a powerful effect on the discussion of the use of money in public education in the United States, not only to detect inequities in the state system, but also to relate dollars to results and the level of efficiency. One might even say that *Serrano* created the need to relate state financing of schools to micro issues of whether money matters at all for children and schools (see Picus & Wattenbarger, 1995; Hanushek, 1986, 1989, 1994).

THIS BOOK

The title of the book, *Optimizing Education Resources*, was selected for several reasons. First, in times of growing student numbers and more needy students, renewed demands for education improvement in a competitive global economy, and declining societal spending on education, we must decide how to "optimize" what we do spend. "Optimizing" is a fairly neutral term: it casts no aspersions on current practices. It simply means to do better with what we've got. Second, we see optimization in four different but closely related ways:

1. To increase *equity* in spending: fairness for youngsters in poor and rich communities, for the able and disabled or challenged pupils, for white, Latino, and African American, and other children of color.
2. To improve the *efficiency* of schools: doing the job of teaching and learning in the best manner with the wisest uses of resources, meaning best results with limited funds.
3. To raise school *productivity*: better results with the most effective uses of resources.
4. To enhance *accountability*: better, more accessible and timely information on school expenditures and results that all stakeholders can share and understand. As schools enter the Information Age, with models and methods elaborated so well in this volume, educators, decision makers, and consumers alike should learn the "facts of the facts" and act accordingly.

The book has yet another important purpose. Besides examining the problems and needs surrounding the optimization of resources, this book also contains a whole set of new, practical approaches or strategies for improving the use of school resources. Warren Howe in Chapter 2, for example, provides great political insight into the slings and arrows of trying to raise taxes to meet judicial demands for statewide financial equity. Thomas Jones in Chapter 3 and Karen Hawley Miles in Chapter 4 give useful advice on the ways both private (EAI) and public (Boston) systems manage education resources.

Those interested in learning how to analyze the ways funds can be used more equitably should read Deborah A. Verstegen (Chapter 6), Sheree T. Speakman et al. (Chapter 7), Martin Orland and Carol Cohen (Chapter 5), and Linda Hertert (Chapter 11). All these chapters contain original models for analyzing equity that range from the schoolhouse to the statehouse to the U.S. Capitol and back. Richard Rothstein presents arguments and strategies for refuting claims that the United States spends and spends its hard-earned resources on public education and nothing positive happens, while Lawrence Picus (Chapter 9) provides powerful indicators that show the practical results of far-reaching state laws for interdistrict equity across Texas, where ironically the original *Rodriguez* (1973) decision stalled momentarily the march to fairness in education funding. "Leveling down" works—as Governor Florio had attempted in New Jersey (see Howe's Chapter 2).

Finally, Dominic Brewer and Dan Goldhaber (Chapter 10) reframe the way we study teacher resources, giving special credence to the "unobservable" characteristics of teacher motivation, classroom practices, and student peer behavior. This formulation should be useful in every district: focusing attention on teacher training, drive, knowledge and skills, along with school funding and policies.

And these practical examples only scratch the surface, identifying the wealth of tools, concepts, and methods in these chapters. This book is both a primer to explain the uses of resources for education and a sophisticated advanced seminar for developing and applying new measures of resource equity and productivity. Thus, this volume is testimony to the rapid maturing of a key field of inquiry, presenting new ways of exploring and improving education structures, funding, equity, privatization, politics, and reform.

This book, therefore, looks comprehensively at and beyond the effects of *Serrano* (1971) after a quarter century. Has the legal avenue worked to produce greater education equity, quality, and accountability in the United States? How is financial equity related to issues of school efficiency, quality, productivity, and accountability? For equity is only part of the answer. What if a nation's schools were all perfectly equitable with resources being spent equally by category of pupil? But these same schools were also incredibly inefficient, downright wasteful, and of terribly poor quality. These schools, while equal to one another (and thus equitable in a sense of "treating all similar children similarly"), were as a group spending their money on trivial, unimportant materials and programs and were so costly to run that all children were equitably deprived. How "fair" is this?

The book is divided into three sections: Part I (Chapters 2-4), titled "Case Studies of School Equity and Effectiveness: New Jersey, Hartford CT, and Boston," provides in-depth pictures of a variety of finance issues, including Chapter 2 by Warren P. Howe, the impact of court-ordered equity in New Jersey; Chapter 3 by Thomas H. Jones which analyzes the privatization efforts in New Haven, Connecticut; and Chapter 4 by Karen Hawley Miles that performs an analysis of school organization and structure and their impact on spending in the Boston Public Schools.

Part II, "The Battle for Education Equity" (Chapters 5-7), focuses particularly on equity in school spending at three different levels in the school system: Chapter 5 by Martin E. Orland and Carol Cohen seeks to explain the effects of federal cutbacks and block grants on state and district spending; Chapter 6 by Deborah A. Verstegen presents new ways of measuring equity based on an eight-state comparative analysis while Chapter 7 by Sheree T. Speakman and colleagues traces costs from the district level to schools and classrooms.

Part III, "Improving School Productivity: From Resources to Teachers to Student Learning" (Chapters 8-11) includes a critique of the "more money/ less learning" argument that has characterized some of the finance and productivity research of late (Chapter 8 by Richard Rothstein). In Chapter 9 Lawrence O. Picus traces the effects of changes in Texas laws on equity particularly since the state's equalization strategy has been to "level down"—cutting the ability of wealthier districts to maintain spending while encouraging and buttressing the low-wealth districts. Chapter 10 by Dominic J. Brewer and Dan D. Goldhaber analyzes the effect of teachers as the main resource (and often the greatest collective cost) on student learning. And Chapter 11 by Linda Hertert examines the relationship between key district demographic variables and spending levels as a means for unpacking the complex interaction of school district efficiency and equity.

PART I: CASE STUDIES IN EQUITY AND EFFICIENCY

In particular, this first section of the book starts with Chapter 2 by Warren P. Howe (U.S. Military Academy at West Point and Fordham University), that provides a political history of recent attempts to follow *Serrano* and *Rodriguez* in New Jersey, and the political "costs" involved for key actors such as Governor James Florio who lost a reelect bid over controversies concerning school equity.

New Jersey has witnessed five *Serrano*-like court cases, starting with *Robinson* v. *Cahill* (1973) and *Abbott* v. *Burke* I (1985), *Abbott* v. *Burke* II (1990), *Abbott* v. Burke III (1994); and *Abbott* v. *Burke* IV (under court deliberation). James J. Florio, previous New Jersey Governor, responded to to the state's highest court by helping to pass an increase in the income tax to fund *Abbott* v. *Burke*. The politics of this legislation caused the first-term New Jersey governor to lose his seat. What were the political factors that surrounded the equity cases in New Jersey and how did the governor get swept up in them? This chapter shows the potent effect of attempted financial equity on the political fortunes of key leaders. Little did the framers of *Serrano* realize just how traumatic the implementation of court rulings on equity can be.

Chapter 3 by Thomas H. Jones (University of Connecticut) is a detailed case analysis of the efforts of Education Alternatives, Incorporated (EAI) to operate the public schools in the Hartford, Connecticut. Jones places the cost effectiveness issue into a "corporate" framework, where the head of EAI, John Gollie, had to manage resources not only to cover the expenses of operating the district's schools for Hartford's 26,000-pupil school system, but also to show a profit, while overcoming the political attacks of the unions and other anti-privatization groups. This chapter places school funding into a new context, while shifting debate to the cost side. In a field where revenues are expenditures, having to contain costs and "show a profit" puts a very different slant on school finance.

Chapter 4 by Karen Hawley Miles (Harvard Graduate School of Education) is a case study of the deployment and use of teachers in the Boston Public

Schools—teacher-related expenditures being the largest single-cost item in education systems. Her analysis exposes the ways in which adding teaching resources to improve "equity" for special categories of students has contributed to the fragmentation and specialization of teaching resources which almost certainly does not lead to the most effective education for all students. Miles begins by exploring the reasons for large class size and unmanageable teaching loads at the high school level in Boston, despite students-teacher ratios that average 17 to 1 nationwide.

Miles finds that four managerial practices describe why students and teachers spend their school days in group sizes far larger than the teaching resources would indicate. These four include: (1) Large number of *specialized teachers* working outside the classroom in "support," "supervisory," "evaluative," and "coordinative" capacities; (2) Short, fragmented teacher planning periods with other teachers "covering," as specified in the union contract; (3) Formulas that group students by formal classifications and contract guidelines, not school-level decisions; and (4) Fragmented daily schedules in high schools with teachers instructing five different student groups daily for less than an hour per class. When Miles combines these school arrangements and structures, they account for 42 percent of lost resources for children's education.

PART II: THE BATTLE FOR FINANCIAL EQUITY: FEDERAL, STATE, LOCAL DISTRICT, AND SCHOOL-SITE

But equity in education is far more complicated than equalizing (overcoming) the effects of local property wealth. It requires an equitable source of resources, means for tracing these resources to students, and legal means for insuring that funds are spent equitably and wisely.

Chapter 5 by Martin E. Orland (National Center for Education Statistics) and Carol E. Cohen (The Finance Project) explores the fiscal future for state and local spending—an especially important topic given the devolution of program and fiscal responsibilities from Washington, DC to states and localities. One of Orland and Cohen's major arguments is that federal cuts in education spending are only a small part of the picture. Under the block grant provision of emerging federal legislation, state and local governments will be increasingly asked to choose between funding public education or other critical needs, such as health and welfare to the poor and elderly. These new pressures, when coupled with anticipated weaker economic growth and school enrollment increases, can be expected to make it extremely difficult for states in general to continue to increase per pupil spending for schools at the same healthy rate as the last two decades. And these conditions also make it most unlikely that the large, long-standing disparities among states in education spending levels will be appreciably diminished in the foreseeable future.

In Chapter 6, Deborah A. Verstegen (University of Virginia) synthesizes and extends advances in the measurement of equity and applies these concepts using an unusually large database of eight states across two school years, employing a wide variety of equity measures. She finds in most states, applying a full range of equity measures, a gradual reduction over time in the "fairness" or equity of education spending. Wealth and school revenue continue to be closely related and appear to be growing yet stronger. Thus, "ability to pay" accounts for half to two-thirds of local expenditures, leading her to conclude that we may be moving away from equality of education opportunity.

Chapter 7 by Sheree T. Speakman and colleagues (Coopers & Lybrand L.L.P.) drives the "fairness" argument to the individual school level, arguing that real equity for children cannot be analyzed using school district data alone. Applying the newly developed Finance Analysis Model™ (developed at Coopers & Lybrand L.L.P. into a software package called In$ite™), Speakman and co-authors were able to trace funds to direct classroom instruction for students for a real but anonymous system, "Mayville Public Schools, USA," by each school. Then applying standard equity measures (e.g., Gini coefficient, McCloone index, coefficient of variation), the researchers determined that greater inequities exist within school systems than between them, a finding confirmed by the work of Linda Hertert in California (Hertert, 1995). The chapter also separates measures of vertical and horizontal equity (see Berne & Stiefel, 1984), again by school site, to show that weighted indexes of need for special populations can be applied using standard formulas.

The next step in Chapter 7 is to determine if school-site financial data can be used to build an equity suit. The authors apply findings on inequalities in individual school spending to the legal precedents based on *Serrano*. The legal argument begins, perhaps, with *Hobson* v. *Hansen*, a 1967 desegregation case in the District of Columbia public schools. Judge J. Skelly Wright of the DC federal court ruled that "Negro" schools were receiving one-third less resources than "white" schools within the same system. This chapter provides the rationale and consequences of applying school-site spending to legal determinations concerning school-level equity.

PART III: IMPROVING SCHOOL PRODUCTIVITY: FROM MONEY TO TEACHERS TO STUDENT LEARNING

The third section of the book moves from equity to productivity: the relationship between school inputs, the processes of teaching and learning, and pupil attainment and growth. Richard Rothstein (Economic Policy Institute) in Chapter 8 argues that the absence of a uniform means for reporting school spending has confounded the relationship between education expenditures and pupil results. He concludes that despite all the arguments to the contrary, the growth in spending is both smaller and more complex than most critics of public education assume. He devotes much energy to the "costs" of special education, and the great difficulty in measuring results. Special education is not a single program but many, with students of varying competencies and problems. Individualization is so critical in special education that "outcome targets" are nearly impossible to obtain, standardize, and apply.

Lawrence O. Picus (University of Southern California) in Chapter 9 traces the effect of changes in state equalization policies on the 1,044 school districts in Texas where legislation (Senate Bill 7 and Senate Bill 351) sought to reduce the spending of rich districts and increase resources for the poor ones. This "leveling down" approach was tested statistically by Picus. In a series of equity measures, he shows the positive effects of equalized tax rates and state aid formulas on education equity between 1988 and 1993, using weighted slopes, elasticity, and various measures of range, dispersion, and equity.

Dominic J. Brewer (Rand Corporation) and Dan D. Goldhaber (CNA Corporation) in Chapter 10 reexamine the long-standing problem of relating resources to results in education, particularly given the widespread belief that money doesn't matter and that basic changes in the use of resources (making classes smaller and investing in teacher training) have little positive (measurable) effect. The authors take the 10th-grade math score results from NELS 1988 study (including a sample of 5,149 students, 638 schools, and 2,245 mathematics teachers) and relate them to teacher characteristics. The analyses imply that some teacher behaviors do affect student test scores, as do teacher characteristics. In particular, the authors find that math-specific teacher training positively affects students' mathematics test scores. This finding is potentially important for policymakers and schools administrators.

Linda Hertert (Education Commission of the States) examines another long-standing problem, that efficiency and equity in school spending are antithetical concepts. Instead, Hertert's Chapter 11 shows the importance of considering local economic variables, such as income levels per capita, which is a more convincing indicator of local school spending than the number of minority pupils. Thus, as states move to increase "power equalization," that is, to remove the link between local fiscal capacity and revenues received and expended on schools, we see that efficiency locally and the ability of states to enforce equity in spending are in fact related.

THEMES AND CONTRIBUTIONS

These following ten chapters—divided into three thematic sections—provide a number of noteworthy common concepts and original contributions to research on and practices of school finance in the United States.

Time Series Studies

Several chapters examine financial changes and equity "over time," an important development in understanding the effects of policies on funds and resources on learning. Deborah Verstegen in Chapter 7, for example, analyzes both between-state and within state variables over time and finds an increasingly nonwealth neutral distribution and a decrease in equity, both between and within the eight states analyzed.

These time-series findings by Verstegen are strikingly different from those of Lawrence Picus (Chapter 9) in Texas, where legislation appears to have dramatically improved equity in spending on education between the 1992-1993 and 1993-1994 school years. Both Picus and Verstegen's research reinforces the need for cross-jurisdictional (state, district, and school) and longitudinal approaches to strengthen the explanatory power of research models.

School-Site and Classroom Analysis

Another new area of study is the school site, where presumably children are taught and learning occurs. Miles in Chapter 4 examines the teacher and classroom organization and the impact on resources, while Speakman and colleagues in Chapter 7 provide data from a new model that moves analysis and reporting to the school and classroom. Both chapters illustrate the incredible analytical power of doing interschool and intra-school studies, over more traditional studies of district-level spending and equity. These practices are promising as school-site management and decision making take hold all across the nation.

Multiple Measures and New Methodologies

This book also illustrate the high level of original work going on in the field of school finance. For example, Deborah Verstegen (Chapter 6) has created a new equity measure, one that we propose be called the "Verstegen Index." It measures equity for those districts (and even schools) that spend in the upper half of the distribution, much as the McLoone Index analyzes the lower half. The Verstegen Index decreases as equity increases across time and states. Verstegen found, using her new index, that the states of Alabama, Missouri, Ohio, and Virginia had greater dispersions in the top half of the distribution over time. This means greater inequalities among those districts with above-the-median spending on education.

Chapter 7 by Speakman and colleagues presents the results of a new school-site Finance Analysis Model™, and the software that supports it, called In$ite™. For the first time, school districts can cross-walk and allocate their district chart of accounts and spending electronically into an analysis methodology that reports 100 percent of expenditures by *function* (e.g., instruction, instructional support, operations, and leadership), *program* (e.g., special education, bilingual/LEP, Title I, general education, and other programs), and *location* (e.g., school-site, central office, and non-allocated). This new tool makes possible school-site managerial reporting, school-site equity and productivity analysis, and the basis for applying the *Serrano* arguments to inequities in school expenditures.

And the data sources in this book range from a single, intensive case study of one state, New Jersey, (Chapter 2, Warren Howe) through school district information and analysis, to some of the largest data sets in the world. For example, Hertert's research (Chapter 11) is based on two national databases, the Common Core of Data (CCD) from the National Center for Education Statistics and the Census School District Special Tabulation of the U.S. Bureau of the Census and NCES.

Tracking Policy Changes

This book also provides research and lessons on the effects of change in finance policies on education politics, quality, and equity. In Chapter 2, for example, Howe analyzes the four-year political effects of increases in school equity funding on the politics and reelection of a once-popular governor, James J. Florio. In Chapter 5, Orland and Cohen follow the effects of federal cutbacks—coupled with federal block grants to states for social, medical, and educational services—on school finance. Orland and Cohen warn of the impact of shifting responsibility to ill-prepared states on the future investments in education. This chapter permits us to trace the effects of changes in one sphere of spending on another—over a number of years. Stresses at one level (federal) may exacerbate cutbacks at other levels (state and local), as communities face rising pupil demographics and need, and the education enterprise becomes vulnerable to cuts in funding.

Jones in Chapter 3 details the "cost effectiveness" of outsourcing education to a private, for-profit company (Education Alternative, Inc.), while Picus (Chapter 9) examines the increase in financial equity resulting from strong legislative action in Texas. And, in contrast, Verstegen's research generally illustrates an "erosion" of equity occurring in the United States in general and in particular in the seven out of the eight states she analyzed. These chapters underline the importance of linking changes in expenditures to policy, program, and equity.

Impact of Conflicting or Competing Goals, Policies, and Research

Finally, this book shows just how complex the financing of American schools can be. Picus in Chapter 9 noted that when the Texas legislature attempted to adjust funding upward for smaller districts (those with fewer than 500 pupils), the policy actually undermined another state policy: to increase equity, since many of the smaller districts were relatively high spending.

Brewer and Goldhaber in Chapter 10 note the confusion and controversy over whether "money matters" in explaining pupil achievement and thus school productivity. These authors argue that part of the problem with school productivity research to date—around resources, teacher skills, and qualifications (as predictors of pupil attainment)—is the over attention to "observable" teacher characteristics such as experience and training, per pupil expenditures, and class size (the Z_1 variables), and the apparent omission of the "unobservable" variables (Z_2) like teacher skills, motivation, behavior, and peer effects.

Using data from the National Education Longitudinal Study, NELS 1988 and 1990, Brewer and Goldhaber found in testing 8th- and 12th-grade mathematics students a "value added" effect or "gain score production" based on teacher behaviors and qualifications. For example, the authors report that pupils having teachers with little or no control over their teaching techniques also have significantly lower test results. Also, in contrast, teachers trained specifically in mathematics have students with better test results. In all, these chapters show the importance of resources, financial and human, in shaping the school environment, setting policies and practices, and attempting to improve learning.

In Chapter 11, Linda Hertert informs our discussions about improving the distribution of education resources (equity) by considering the relationship between the variation in district revenues across the country and certain key school district characteristics. Findings indicate that (1) the percent of minority pupils in a district is not a reliable indicator of revenues; (2) poverty is not solely an inner-city issue; and (3) per capita income, of the variables considered, has the most consistent positive relationship to per pupil revenues. When 1989-1990 district revenues were adjusted according to the variables considered, the greatest improvement in equity was achieved using the income factor.

SUMMARY

This book illustrates the complexity of the impact of *Serrano* over the last 25 years. Starting with simple interdistrict per pupil spending, we have moved to highly political effects (Howe, Chapter 2), to attempts to surpass the strictly "public" view of education toward privatization of schooling (Jones, Chapter 3), and to internal structural effects on spending (Miles, Chapter 4), as these three case studies (Part I) indicate.

Equity is more than an interdistrict phenomenon. Orland and Cohen (Chapter 5) indicate the effects of federal cuts and block grants on state and local spending on all social services. Speakman, Cooper, Bloomfield, and associates (Chapter 7) analyze school-site expenditures and their levels of both horizontal and vertical equity—a move far beyond what the *Serrano* architects could have imagined in 1971. These authors also speculate about a new kind of court case involving intra-district equity and adequacy, exploring what might happen if *Serrano* did indeed "go to school."

And equity inevitably leads to productivity analysis: for it is difficult to sustain a campaign for "fairness" of financial inputs if this spending cannot be related in some way to student learning and development. Rothstein (Chapter 8) attacks the attackers of public education who assert that we are spending more for less results. He explains that we do not know how to compare from year to year, region to region, given the idiosyncrasies in categories, spending, and students. Picus (Chapter 9) and Verstegen (Chapter 6) apply methods of measuring education equity to states: Picus in Texas and Verstegen in eight others. And they conclude very different things: Picus found increasing equity in Texas as the legislature passed laws "leveling down" the high-spending districts while raising up the poorer ones; but Verstegen's picture is somewhat bleaker, as her time-series analysis detects "erosion" in equity in most states.

Chapter 10 by Brewer and Goldhaber applies some basic correlation models to the relationship between pupil characteristics, teacher variables, and student learning outcomes (NELS, 1988). Hertert in Chapter 11 relates local demographic conditions to the spending of districts using a national data set, dispelling some of the notions of race and minority status as the best predictors of education spending. These chapters again illustrate the importance of accurate information on schools, teachers, students, and their programs—if we are to make real advances in education productivity in the United States.

In conclusion, this volume shows the historical, social, economic, and methodological problems in determining whether *Serrano* (both as a reality and a symbol of equity) has spawned a movement toward better schools, better financing systems, improved teaching, and better paid and prepared staff, and any real improvement in student learning. Arthur Wise, in his book *Rich Schools, Poor Schools*, recognized six years before *Serrano* the incredible complexity of school productivity. In fact, even though these words were penned 28 years ago, they sound as relevant now as then, and are a fitting focus for the book. Wise wrote:

> Determination of the "acceptable norms," however, cannot depend on some fixed ideal. In effect, educators are viewed as setting norms for all students at a level which will just exhaust the available educational resources. Such resources, like all resources, can be assumed to be in limited supply. Since the educators' task is, therefore, to determine how the available resources are to be allocated, and since, clearly meeting all the identifiable, ideal needs of all students is impossible, the educator must set levels of expected achievement for all students in such a way that the *available* resources are exhausted. Such levels can

be viewed as the acceptable norms, to be used in turn for determining each youngster's education needs (Wise, 1967, p. 145).

This argument sounds amazingly contemporary: how does one devote available resources to meeting the needs of all our students in ways that surpass what is now "expected" but are still affordable, given the limited supply of resources and the mounting needs and numbers of students? Surely, increasing equity, strengthening our knowledge of costs and financial and education practices, and relating these factors to student development are the mission of this book and a critical concern of education and society.

REFERENCES

Berne, R., & Stiefel, L. (1984). *The measurement of equity in school finance.* Baltimore: Johns Hopkins University Press.

Carrington, P.D. (1973). Equal justice under law and school finance. In *School finance in transition.* Gainsville, FL: National Conference on School Finance.

Hanushek, E.A. (1986). The economics of schooling: Production and efficiency in public schools. *Journal of Economic Literature, 24*, 1141-1177.

———. (1989). The impact of differential expenditures on school performance. *Educational Researcher, 18* (4), 44-51.

———. (1994). *Making schools work: Improving performance and controlling costs.* Washington, DC: The Brookings Institution.

Hertert, L. (1995). Does equal funding for districts mean equal funding for classroom students? Evidence from California. In L.O. Picus & J.L. Wattenbarger (Eds.), *Where does the money go? Resource allocation in elementary and secondary schools* (pp. 71-84). Thousand Oaks, CA: Corwin Press.

Picus, L.O., & Wattenbarger, J.L. (Eds.) (1995). *Where does the money go? Resource allocation in elementary and secondary schools.* Thousand Oaks, CA: Corwin Press and the American Education Finance Association.

Wise, A.E. (1967). *Rich schools, poor schools: The promise of equal education opportunity.* Chicago: University of Chicago Press.

PART I

CASE STUDIES OF SCHOOL EQUITY AND EFFECTIVENESS: NEW JERSEY, HARTFORD, CT, AND BOSTON

Chapter 2

THE QUEST FOR EDUCATION EQUITY IN NEW JERSEY: THE EXPERIENCE OF GOVERNOR JAMES J. FLORIO

Warren P. Howe

INTRODUCTION

New Jersey is a useful case for several reasons. First, it has been among the most active states in the post-*Serrano* era with four major court rulings on educational finance equity since 1973, beginning with *Robinson* v. *Cahill*, 62 N.J. 473 (1973). A series of *Abbott* v. *Burke* decisions followed with the first coming in 1985 (100 N.J. 269). *Abbott II* (119 N.J. 287), in June 1990, was the catalyst for New Jersey's equity legislation, the Quality Education Act of 1990 (QEA I), and *Abbott III* (136 N.J. 444), which ruled a 1991 amended QEA unconstitutional, came in July 1994. As this paper is being prepared, a fourth *Abbott* litigation, initiated in April 1996, is awaiting court action.

Advances in Educational Productivity, Volume 6, pages 17-42.

ISBN: 0-7623-0082-5

Second, the political history of the *Robinson* decision and its aftermath are well documented by Richard Lehne (1978) in his book, *The Quest for Justice*. This study helps fill a research void on school finance equity, detailing the background and political repercussions of New Jersey's first equity suit. Lehne explains what happened once the state supreme court declared New Jersey's school finance system illegal under the "thorough and efficient education for all" language of the *New Jersey Constitution* (1947).

And third, the fate of Democratic Governor James J. Florio became intertwined with efforts to increase taxes and extend greater interdistrict equity. With a Democratic majority in both houses of the New Jersey Legislature, Florio was able in just six months to resolve a recession-induced budget deficit inherited from the previous administration of Governor Thomas Kean, legislate a redistribution of state resources for a more equitable financial base among school districts, and raise the necessary funds by restructuring the teacher pension system, and by increasing state sales and upper-level income taxes, increases which would produce 2.8 billion new dollars. He ultimately won the battle but lost the war. This rapid and decisive action cost the Democrats their majority in the Legislature in 1991 and two years later saw Governor Florio unseated in the gubernatorial race by a little-known conservative Republican, Christine Todd Whitman.

THE STUDY

This chapter is a political history of the dynamic Florio years as the Governor struggled to equalize education opportunity in compliance with the mandate of *Abbott* v. *Burke*. Five key questions are considered:

1. What was the political philosophy and follow-up legislative strategy of Governor Florio and Democratic lawmakers as they faced a budget deficit, a court mandate to initiate an education equity plan, and a consequent need for new and higher taxes?
2. What barriers to equity arose as a result of the Florio program?
3. What were the reactions of legislators, educators, interest groups, and New Jersey voters in general to the Florio administration course.
4. What evaluation can we make of the Florio administration and the New Jersey Democratic Party as a result of their battle for education equity financing?
5. What lessons have been learned from the Florio experience which are of value to other education equity challenges in the nation?

THEORETICAL RATIONALE

A political history calls for the careful blend of historiography and political science. Interestingly, analysis of New Jersey can be explained by the classic theory of political behavior advanced by Theodore J. Lowi (1972) which includes his *regulatory*, *distributive*, and *redistributive* political styles. Much education politics seems to fit into the *regulatory* style wherein a few key interest groups, usually those most affected, battle for or against particular laws. Wider participation is often negligible. For example, when the apple industry is confronting new regulations, conflict is normally limited to the principal political stakeholders, supermarkets, and apple growers.

The second style, *distributive* politics, are those policies that grant resources to large, broad constituencies which often close ranks to seek more money. In the politics of Chapter I (*Title I, Education Consolidation & Improvement Act of 1981, P.L. 97-35*, which concerns education aid for disadvantaged children), for example, virtually all teachers, parents, and administrative groups, working with advocates for education, move together for more resources. The third style Lowi calls *redistributive* politics, a form often involving one group gaining resources or opportunities at the expense of other groups.

Typically, education finance has involved distributive politics primarily as federal and state governments provide resources to almost every district, school, and teacher in some form or another. As predicted, teachers' unions, administrator and principals' associations, school board groups, and other boosters of public education unify to get more money for everyone. This battle is often about "how much more," not whether public education should be funded.

But as this chapter illustrates, education equity litigation, which has been New Jersey's circumstance since 1973, was converted by Governor Florio from a distributive to a redistributive political issue—from everyone gaining something, to the poor and middle-income groups receiving from the rich; and from inclusive politics to battles of class and race which divided stakeholders by their wealth and district locations (urban, suburban, or rural). Education went from being a mild, long-term issue to one with "winners" and "losers" based on socioeconomic, racial, and regional lines.

THE GOVERNOR'S PLAN

Opponents of Governor Florio's six-month legislative plan for bringing financial equity to New Jersey's 620 school districts while simultaneously addressing a $600 million budget deficit, correctly termed his program: "too much, too soon." This label, offered as a critique, actually described the intended strategy. The Governor, by design, sought to address many complex issues quickly, while the Democratic legislative majority lasted, risking controversy and hoping that public support could be won before the next election. His agenda was purposely rapid, diverse, and guided by a leadership philosophy, spelled out in the Governor's first Budget Address, of "determining what is politically possible, and then doing more... it's only when you push the edges that you really excel" (Budget Address, Governor James Florio, March 15, 1990).

THE "WINDOW OF OPPORTUNITY"

The Governor and his ally, Democratic Assembly Speaker, Joseph V. Doria (31st District, Bayonne), both believed that the risk was acceptable and that a reasonable chance existed for ultimate political acceptance. Several reasons guided this belief: (1) their long political experience in the state; (2) the long-term pressure of court-mandated fiscal equalization dating from 1973 with *Robinson* v. *Cahill* through 1990 with the second *Abbott* decision; (3) their margin of election victory, and (4) Democratic control of the Legislature.

Political Experience

James Florio had come to the Governor's Mansion directly from the U.S. House of Representatives where he had served the Camden County Congressional District for 12 years after two terms in the State Assembly and service as Camden's assistant city attorney. A native of Brooklyn, New York, he had dropped out of high school, joined the Navy where he earned an equivalency diploma, graduated from Trenton State College under the G.I. Bill, then earned a law degree from Rutgers University. Florio had deep respect for the value of education, especially for children growing up in urban low- and middle-class communities. Before choosing the law as a profession, he had been a Woodrow Wilson fellow at Columbia University, training to be a history teacher. Today, in addition to the practice of law, Governor Florio teaches at Rutgers University.

The other key political actor, Assemblyman Joseph Doria, was beginning his 10th year in the statehouse at Trenton. A native of Bayonne in the urbanized region of northern New Jersey, Doria is an educator by profession. A former high school teacher, he was president of the Bayonne School Board during the mid-1970s and has been a member of the New Jersey Assembly continually since 1979 with considerable service on education committees. In 1990, during legislative consideration of the Florio administration's *Quality Education Act of 1990*, he was Assembly Speaker and now serves as leader of the Assembly Democratic Party.

Court Mandates, Popular Support, Legislative Control

Court decisions, evidence of strong popular support, and a legislative majority also made legislation for increased school aid, especially in poor urban areas, seem possible. The New Jersey Supreme Court, in its 1973 *Robinson* v. *Cahill* decision, had stressed that the state constitutional requirement that all students receive a "thorough and efficient" education, includes the presence of educational opportunity and that sufficient funding is an important part of this opportunity, especially in urban areas. The legislative result was *The Public School Education Act of 1975* and a new state income tax established as the funding vehicle.

In 1990 a second New Jersey Supreme Court decision in *Abbott* v. *Burke*, first initiated in 1981 by the Education Law Center, an urban advocacy group based in Newark, New Jersey, was pending and was expected to require that New Jersey fund its schools on a more equitable basis. The state school aid process then current was based on the tax potential of local property values, and did not include a state required standard of per-pupil spending. Consequently, the disparity in funding between urban school districts and their wealthier suburban counterparts was wide and growing larger, perpetuating the violation of "thorough and efficient." The *Abbott* decision would require a new school aid plan that would close this gap. Hence, the Governor and the Assembly Speaker had the weight of the state supreme court to support their legislative efforts.

In addition to legal justification, the Governor perceived that he had a popular mandate to equalize spending. In the November (1989) election, Governor Florio had defeated his Republican opponent by a "landslide," to use the term of the Newark *Star Ledger*, the state's major newspaper. Florio had in fact won by 62 percent of the vote over rival Congressman James Courter. He captured 36 of the state's 40 legislative districts, while the Democrats gained a majority in both houses of the New Jersey Legislature. Although some Democratic legislators, including Joseph Doria, felt that the Democratic victory had been more the result of mistakes by the Republican candidate than the popularity of the Democratic platform, the Governor's majority, plus Democratic control over both houses, suggested that it was politically possible to gain passage of an equity agenda.

THE OPPOSITION: THE RISK OF CHANGE

With these assets of experience, judicial decision, popular support, and legislative control, the Governor, aided significantly by the legislative skill of Joseph Doria, was able to pass a school equalization law and related tax package by late June 1990. Timing was critical not only because the budget deficit required resolution before the new July 1 fiscal year, but because of the need to utilize present voter and legislative support, always in danger of eroding when controversial goals are at stake. But this rapid, complex legislative achievement did not come without a price. Opposition groups had begun to emerge by the early spring, even before the Florio program was formally submitted to the Legislature. The groups were as follows:

The Voters from Wealthier School Districts

Governor Florio's education equity proposals were classic "redistributive" politics, to use Theodore Lowi's term. But the redistribution gave wealthier school districts a "double punch," not only forcasting eventual reduction or elimination of their state aid (after a transition period) but also forcing their more affluent citizens to pay higher income taxes to fund assistance for lower- and middle-income districts. Taxpayer resistance was demonstrated initially through the amendment process and finally through the ballot.

The Teachers' Union

The Governor's legislation incurred the wrath of the key education interests, including the state's largest teacher's union, the liberal Democratic New Jersey Education Association (NJEA), by requiring that local school districts take over full funding of teacher pensions. Removing this responsibility from the state, with the exception of back-up guarantees, would free substantial funds for state education aid, but would cause fear within the NJEA that the added burden of pension contributions might diminish local funds for teacher pay raises and available resources for classrooms, and might, as a consequence, erode union power to negotiate salaries effectively.

Antitax Groups

Antitax organizations also opposed the Governor because the school aid plan would require an upper-level income tax increase scheduled for passage at the same time as the deficit closing sales tax, which, on July 1, would affect everyone in the state, rich, middle income, or poor. Thus, the Governor managed to incite the rich with his targeted income tax hike, and everyone else with an increase in sales taxes. This double-barreled "package" provoked taxpayers and stimulated organized opposition because of a campaign statement by the Governor (issued before the recession created an unexpected deficit) that he did not expect tax increases would be necessary based on the forecast of a surplus by the outgoing administration of Governor Thomas Kean.

The State and National Gun Lobby

One month before the school aid/tax package was approved, the Legislature had passed an administration-sponsored assault weapons ban. In reaction, state and national gun lobbies became a lucrative ally of the state teachers' union and antitax forces in opposing the Governor. As a result, the substantial resources of the National Rifle Association were added to the coalition of education, antitax, and other groups opposing Florio's reelection.

The emergence of this organized opposition demonstrated the classic political problem faced by the Governor and his strategists: How to move quickly, capitalizing on the "open window" of strong election victory and its consequent legislative expectation, without offending important constituencies, and ultimately losing support from the majority of voters.

POTENTIAL BARRIERS TO EQUITY

The "open window" opportunity, the opposition, and the problems they together posed, were reflections of several issues that became potential "barriers" to the achievement of financial equity in New Jersey. They were:

1. *Recession:* An economic down-turn in New Jersey, part of the national recession during the late 1980s and early 1990s, cut state tax revenue, created a state budget deficit, and required both increased sales taxes to close the financial gap and new income taxes to provide sufficient funds for school equity as required by the court.
2. *Tax "Package":* The combining of the state sales and income tax hikes as a legislative "package" alienated voters and became a visible target for antitax groups.
3. *Campaign Tax Statement:* During the 1989 campaign, the future Governor stated that he saw no need for new taxes based on the forecast of a budget surplus by the outgoing administration of Governor Kean. The tax increase proposal shortly thereafter, in the face of a surprise deficit, was perceived by many voters as a broken "promise."
4. *Methods:* The exclusive, almost secretive way in which the Florio administration drafted the education aid bill, alienated important members of the New Jersey education community.
5. *Rapidity:* The fast-paced submission, review, and passage of a complex, far-reaching education bill prevented adequate time for comment and the building of broad-based support.
6. *Pension Plan:* The Florio school aid plan restructured the teachers' pension program, moving payment responsibility from the state to local school districts, thereby increasing available state school aid revenue.

But opponents saw the plan as adding to the burden on local resources, potentially diverting funds otherwise available for teacher pay, benefits, and classroom use. The plan also gave the teachers' union an argument to build opposition by claiming pensions, dependent on financially stressed districts, were in jeopardy.

7. *"Redistribution" of Wealth:* The Florio school aid plan, redistributed resources from relatively wealthy school districts to middle-income and poorer districts, forecasted the eventual reduction or elimination of state aid to the former, and raised upper-level income taxes as well.
8. *Inner-city School District Management:* Many suburban middle- and upper-income voters came to believe that more resources for city school districts were a waste, based on voter perceptions of urban mismanagement—the so-called "bottomless pit" syndrome.
9. *Linkages and Coalitions:* Finally, this panoply of concerns solidified, creating linkages among groups opposing Governor Florio for different reasons (for example, for taxation, pension restructuring, banning assault rifles) thereby enabling the development of a wide-spread, broad-based, highly organized, and well-funded coalition against the Governor's reelection.

The Deficit, Taxes, and Strategy

The first and most significant barrier to Governor Florio's equity plans had been looming across the nation since the stock market collapse of October 19, 1987, "Black Monday," and the savings and loan failures. Like the rest of the United States, New Jersey had entered a recession and outgoing Governor Kean's anticipated surplus of funds for the new governor changed to a deficit by the time of Florio's inauguration. Two months later, as Governor Florio was preparing his first budget address, the state deficit was approaching $590 million. It was obvious to all, even to the Republican opposition, that additional funds were needed to balance the state's 1990 budget, starting July 1. Not only was a balanced budget the law, but the state's triple-A bond rating would be jeopardized if the shortfall was not closed.

Budget cuts could not raise the needed revenue in time; hence, some tax increases were required. This placed Florio in potential trouble with voters. During his campaign, Florio stated that based on outgoing Governor Kean's anticipated surplus, he "saw no need" for a tax increase. This statement was interpreted by many voters as a pledge not to raise taxes. In addition, the Republican opposition candidate, James Courter, had charged during the campaign that if elected, Florio would raise taxes. The Governor was now in a bind. The flagging economy would force him to raise taxes, and former Governor Kean's optimistic prediction had induced Florio to appear as if he were going back on his "perceived" word. The "honeymoon" with the electorate was in danger. Much would depend on the Governor's March 15th budget address.

In his first budget address, the Governor announced the unanticipated deficit and a consequent "harsh" budget which, he admitted, was unfair but unavoidable. To eliminate the deficit, he introduced a one-cent sales tax increase on certain "nonessential" items such as paper goods and household products; food and clothing were excepted. The sales tax, to take effect July 1, would affect all classes of citizens: rich, middle income, and poor alike. Although unpopular, it was not difficult to understand given the budget shortfall.

But the Governor was also commited to assisting education in poor and middle-income areas by achieving statewide equity, especially in urban districts, in accord with the expected mandate of the pending *Abbott* decision. Helping education would require an equal "foundation" expenditure of dollars for every public school student in the state—demanding yet more revenue. Unfortunately for Florio, the proposed source, also announced in the budget address, would be an income tax hike for upper-income families. The Governor proposed an increase for family incomes of $55,000 or more annually. Subsequent bargaining between the Governor and Democratic legislative leadership produced a revised proposal, ultimately adopted by the Legislature in June, which increased the tax for single filers earning over $35,000 and joint filers and single heads of households earning over $70,000 per year. A top rate of 7 percent was set for singles earning over $75,000 and for joint and single heads of households earning more than $150,000.

Initially, the Governor had sought to avoid placing the sales tax and a school aid-inspired income tax increase in the same "package" plan. He had considered delaying the school aid legislation as well as its funding vehicle, the income tax increase proposal, until the end of the year. The court would require a new school aid funding mechanism but not until the 1991-1992 school year (*Abbott* v. *Burke*, p. 389). It was already too late to change the law for the 1990-1991 school year and the deficit had frozen education aid for the current academic term (Goertz, 1992, p. 9). Enactment need not come so early in 1990. Action on school aid and its income tax funding vehicle could wait until January, an adequate date for the plan to be legislatively in place (Corcoran & Scovronick, 1994, p. 16). Delaying consideration of the issue would separate the school aid/income tax subject from the sales tax issue thus making the former more politically acceptable and providing more time to explain state aid to education, a complex issue, to the public.

Thomas Corcoran, the Governor's education advisor, and Nathan Scovronick, treasury department executive, write that later passage of the school aid/income tax proposal would "give the Governor the opportunity to explain the need for each element of the program." Conveying messages in New Jersey, they state, "always requires a lot of time because of the absence of any statewide newspapers or television stations that are widely or regularly watched" (Corcoran & Scovronick, p. 16). The supreme court, they continue, would also have the time to relate the consequences of noncompliance. The Florio administration felt that when the *Abbott* decision came, the court might assist the legislative effort, as it had done in 1976, by threatening a school closure if the Legislature did not provide sufficient equity funding (Scovronick, phone interview, 1996). This would reduce any political burden the Governor and the Democratic legislators might experience from the NJEA or voters concerned about income tax increases.

But Joseph Doria and Assembly Democrats caused a change of strategy. They opposed the sales tax altogether as the least democratic method of resolving the deficit and felt it would present an undue burden on middle- and lower-income groups. They believed that the income tax should be the funding vehicle for deficit reduction as well as school aid (Doria interview, 1996).

The Florio team, however, was unwilling to sacrifice the sales tax. They insisted on the income/sales-two tax strategy. Doria and Assembly Democrats responded with the requirement that if two tax increases were to be, the Legislature would devote only one session to a vote. Florio would not be able to delay the school aid/income tax proposal until later in the year and still expect approval of a deficit ending sales tax before the new July 1 fiscal year. In addition, the Doria group insisted, the income tax must also generate property tax relief.

The Florio team, seeing the need to satisfy the Legislature or accomplish nothing, was forced to combine the income tax and sales tax proposal in a single session "package." The Governor's main objective was passage of school aid. If passage by the Legislature meant voting the two taxes in one early session, then that would be the plan. In a 1995 interview, Governor Florio indicated he was "not uncomfortable with combining the sales tax and income tax in a single package." In these early days he felt the essential goal was to have a program ready for the 1991 school year. And in spite of the court's willingness to permit a "phased in" equity program, the Governor and the Legislature also felt that it was important to have the dollars to effect the full program without delay if possible (Florio interview, 1995).

Doria's resistence to more than one legislative session for tax consideration was partly inspired by the Senate's lukewarm support for the Governor's program. The Senate and its Democratic majority leader, John Lynch (17th District-Middlesex and Somerset Counties), a former mayor of New Brunswick who held state aid to municipal government as a priority, suspected waste in the spending record of school districts, be they urban or suburban. Lynch's instincts called for cutbacks in funds for schools.

Doria, therefore, saw too great a risk in leaving the school aid/income tax funding vote for later in the year after a sales tax increase had already been approved. There would be no assurance of Senate approval (Teele & Muller interview, 1995). The Governor's post election strength might also diminish as the year progressed and Doria did not want to give the Senate the political rationale of weakened public support as reason to oppose end of year school aid/income tax bills.

In addition, Senate Democrats felt, as did the Governor in his original plan, that the *Abbott* court would be flexible with implementation time. They had, after all, specified a "phased in" approach if desired (*Abbott* v. *Burke*, p. 389). They would not expect complete equity in the first year as the final administration Assembly Democrat proposal would ultimately suggest (Crotty interview, 1995). This belief could remove the sense of urgency to approve a school aid plan and a funding source in 1990.

Doria was ultimately successful in gaining the support of Senator Lynch and the Senate Democrats, although it was reluctant support from many Senate members. They would view the Assembly position in much the same way as did the Florio administration, going along with the Doria single session tax vote plan as an alternative to accomplishing nothing (Corcoran & Scovronick, 1994, p. 16).

Corcoran and Scovronick suggest that Assembly Democrats, in addition to their desire to commit the Senate to an early tax package vote, also "wanted a minimum amount of time for legislative controversy and a maximum amount of time before the next election to convince voters of the merits of their position" (Corcoran & Scovronick, 1994. p. 16). They did not want to stretch out the tax increase issue, always politically controversial, over a full year especially when the electorate had been led to believe during the campaign that tax increases would probably be unnecessary. Two legislative seats, that of 1st District Senator James Hurley who had resigned to join the Casino Control Commission, and a 35th District (Paterson) Assembly seat, vacated when John Girenti moved to the Senate at the death of Senator Frank Graves, were scheduled for the off-year 1990 election. The two seats would be important tests for the Democratic agenda. And the general election in 1991 would test voter support for the entire Legislature.

Because of the tenuous position of Senate Democrats and the possible erosion of public support as the year progressed, Assembly Democrats saw a single session and early tax package vote as the best "window of opportunity." The early voting date was essentially predetermined by the coming new fiscal year which would begin on July 1. The voting would need to occur by June 30 so that the deficit would be resolved by the sales tax increase before the new budget took effect. State budgets, by law, may not carry over deficits from previous years. Considering that the Democrats had only been in power since January 1, June was an extremely short legislative schedule. And the tax vote would not be the only controversial issue on the agenda.

By implication, the Assembly single session tax vote requirement meant that the complex school aid bill, the eventual *Quality Education Act of 1990* (*QEA*), would also need to be considered in that session. It would make no sense to vote an income tax increase to fund a school aid law which did not yet exist. There was no consideration, in other words, given to voting the QEA alone later in the year apart from the tax package vote (Scovronick telephone interview, 1996). The school aid vote, therefore, would also occur by June 30.

Inclusion of the school aid bill in the same Legislature session as the tax package had the effect of crowding very complex proposals into an extremely short time frame, leaving Assembly and Senate members little if any opportunity for understanding details and implications. Doria and the Legislature Democrats, in fact, believed the whole program was done too quickly, but they went along with the crowded Florio agenda in the spirit of party unity (Doria interview, 1996), chosing to go for it all in a short time before the "window of opportunity" closed (Crotty interview, 1995), albeit with a reluctance that sensed serious negative political implications.

A variety of compromise decisions, therefore, made by the Florio administration and Democratic Assembly leaders, created a complicated and extensive June legislative agenda that would be difficult to understand and support, not only for the Assembly and Senate, but for the public. The political liabilities that followed might have been avoided if either the Governor's original plans, or the desires of the Doria Assembly group had prevailed unchanged. Ironically, it was the compromise of the two plans that brought the "too much, too soon" legislative agenda of June 1990. The basic democratic practice of concession led to a politically hazardous strategy that may not have made defeat inevitable, but which certainly contributed to its strong possibility.

The total tax increase would produce $2.8 billion, of which $1.5 billion would be directed to school equalization aid. And because the city school districts had the heaviest need for equalization funds, the effect of the Governor Florio's plan, in the minds of many voters, was a redistribution of wealth from the more affluent suburbs to the urban core. Not only were the state's electorate being asked to pay a sales tax increase but also an upper-level income tax hike, the benefits of which appeared to be focused primarily on the urban and rural needy at the expense of suburban children. And all of this came from a governor who stated prior to election that he saw no need to raise taxes at all.

Public response to the budget message was overwhelmingly negative. Citizens did not seem to hear the Governor's entire proposal: that the income tax increase was targeted only at the upper-income taxpayer and that his plan for education improvement would benefit not only the poor districts (mainly rural and urban) but the middle-class school systems as well. The latter was beyond what the *Abbott* remedy would require; but the Florio administration believed that it was only fair and politically acceptable policy. The catch was that inclusion of middle-income districts meant that more revenues were needed.

To counter the negative effect of tax increases, the Governor announced an increase in property tax rebates for low- and middle-income homeowners. Rebates, after all, were funded by the income tax. Since the property tax was the main source of local revenue for schools, the "homestead" rebate was, in effect, an additional state education assistance program. The potential for positive reaction to the rebate increase proposal, however, may have been lost when Florio announced his intention to suspend the program for 1991 in view of the deficit and the need for higher school revenues.

The strong negative public reaction to these proposals appears to have been more a reaction to timing and taxation method rather than to the question of actual need for added revenue. By presenting the entire tax increase program in the first budget message, during a recession, and only a few months after his election campaign when he suggested no new taxes appeared necessary, and by presenting the income tax idea within the context of a controversial school aid redistribution of wealth, the Governor created a coalescence of strong opposition.

Governor Florio had expected some opposition, but felt that the critical nature of the time demanded risk—going beyond what was politically possible. He challenged the state to strive for excellence by supporting a program of short-term austerity for a longer-term gain. Perhaps the public could be convinced in later weeks; but for now the honeymoon was suddenly over.

Isolation, Pensions, and the NJEA

As if the recession, deficit, and tax package were not enough, Governor Florio faced additional problems from the rather closed method his administration employed in drafting the school aid legislation. The bill was drafted by a small group within the executive staff, headed by Douglas Berman, State Treasurer and former Florio campaign manager. Berman conducted a process that excluded formal participation by the education community and other key groups outside the executive branch, although there was informal communication with these interests. This exclusive process of bill-writing did allow for a well-conceived law for structural change, a feat that would have been more difficult to achieve with a larger, more inclusive, and diverse committee. But the relative isolation from the larger community did nothing to build political support for the bill's ultimate public acceptance.

One significant example was loss of support from the major teachers' union, the powerful and usually loyal (to Democrats) New Jersey Education Association (NJEA). Exclusion of the NJEA from formal participation in drafting the school aid legislation occurred over an administration proposal concerning teacher pensions.

The proposal called for the transfer of responsibility for payment of teacher pensions and social security from the state to local school districts. Local responsibility for pensions would free millions of state dollars which could then augment school foundation aid sufficiently to assist not only poorer urban and rural districts, but also middle-income districts. There would also be sufficient foundation aid to provide school districts eligible for aid, the funding assistance they might need to pay pensions. The transfer idea also had a logical basis. Since teachers' salaries and other benefits (health and unemployment insurance) are bargained locally in New Jersey, the school bill reasoned that pensions—which are based on salaries—should also be funded locally, albeit with state back-up guarantees.

The NJEA strongly opposed the pension shift. The association felt the proposal would reduce the power of the union to negotiate salary increases since school boards, facing the additional costs of pensions, would be less likely to bargain higher teachers' salaries. In effect, the NJEA argued that transferring pension costs to districts would divert funds from the classroom (Corona interview, 1995). Berman's bill-drafting group was aware of the NJEA feelings from informal meetings. On one occasion the union even met with the Governor. The administration had no doubt that if the union was involved in the drafting process, they would not be willing to compromise on the pension issue. The NJEA consequently became an additional opponent of the Governor and his school aid legislation—for pension rather than for tax reasons.

The teachers' union advised its membership that the Florio administration was undermining the security of their pensions, and many teachers listened, believed, and organized strong protests. In the mind of the Governor, the pension system was never really at risk, given the state back-up. The Governor and his administration, however, had little if any success in countering the charges of the NJEA. Apparently, more time was necessary to convince teachers of the merits of the district-based pensions. Meanwhile, opposition to Florio was increasing.

Legislator Concerns

By May 1990 the school aid bill was close to completion and ready for submission to the Legislature. The education staff of the Democratic legislative majority had been included in discussions with the Berman drafting group in April and general consensus had been reached. On May 24, the bill was introduced to the Assembly and sent to committee. A similar bill went to the Senate and into committee on June 4—the day before the New Jersey Supreme Court rendered its decision in *Abbott* v. *Burke*. Timing problems reached a crisis point during this legislative phase as the Governor and Democratic Assembly and Senate leadership struggled to muster sufficient support to pass the bill before the end of the month. And the school aid bill with all its complexities was not the only consideration; the controversial package of income and sales tax increases was also on the agenda.

The Governor did not have complete support from the Legislature, even among the Democratic majority. For lawmakers unfamiliar with the intricacies of school aid and double tax legislation, this reluctance was understandable. Most legislators did not see the bill until it was released from committee in mid-June, and the schedule called for a vote the following week, hardly enough time, in the minds of many, to understand fully the complex details and consequences of statewide equity in a state as economically diverse as New Jersey.

Concerns of the Wealthy Districts

Amendments were made to the school aid bill once the draft reached the Legislature. Minor modifications accommodated the *Abbott* v. *Burke* decision, and concerns of wealthier constituents, especially the influencial Bergen County Democratic delegation, were addressed. Although each district in the state was to receive at least as much state school aid for the 1991-1992 school year as it received in the 1990 school year as increased by 6.5 percent, and transition aid was to be paid through the 1994-1995 school year to relatively affluent districts earmarked for eventual termination of foundation aid, wealthier districts still became centers of growing opposition because they faced not only an eventual reduction or loss of aid under the redistribution plan, but also faced higher income taxes. In addition, they would watch their higher tax revenues pass to the middle-income and poorer urban and rural districts. Their own school districts would ultimately depend exclusively on local resources and would face expenditure "caps" to keep their school authorizations in relative balance with the rest of the state. These prospects seemed to threaten their schools with an erosion of excellence, a "snuffing the lighthouse" effect. Additionally, teachers in wealthier districts feared not only the insecurity of their pensions, but also reduced salaries (which also reduced their pensions).

It was not that for affluent areas like Bergen County objected to the improvement of education quality and opportunity for inner cities such as Newark and Paterson, but they saw themselves as paying the bill while at the same time suffering real losses in state aid, personal income, education quality, and retirement security for their teachers. The Governor's homestead tax rebate plan, and the teachers' pension guarantee were somehow not visible in June 1990 to appease the electorate. Something additional was necessary.

Additional concessions came in the form of amendments to separate categorical aid for "at risk" and bilingual students from the basic foundation aid formula. Wealthy districts would now receive categorical dollars on a per pupil basis without regard to the needs criteria of the foundation program. Student transportation aid was also adjusted so districts such as Bergen County, where the population is geographically dispersed, would receive more state aid than the urban school districts where travel is not as extensive or expensive. But amendments to appease the wealthier, northern New Jersey districts made matters still more complicated without adding time for legislative review. Timing again was becoming a increasingly critical factor.

Urban "Competence"

An additional concern was whether poorer "special needs" districts, of which the Florio school aid bill had designated 30 (one more than the *Abbott* decision had specified), would be able to properly manage large amounts of new state aid, in spite of amendments requiring state monitoring and approval of spending plans. This uneasiness (the "bottomless pit" theory) was compounded by the fact that state aid funds were to be "taken away" or "redistributed" from wealthier districts. Further, the full amount of "foundation" aid was to be given to special needs districts for the 1991-1992 school year, while the *Abbott* decision had authorized (but not required) a more gradual or "phase-in" approach.

The Marathon Session: The Partisan Vote

In strict partisan voting, the New Jersey Legislature approved Governor Florio's school aid plan as the Quality Education Act of 1990 (QEA I) during the week of June 17. Although the Senate approved the sales tax, school aid plan, and its income tax funding plan in a one-day sequence, Wednesday, June 20, the Assembly's passage extended to a four-day marathon, with income tax approval on Monday night, approval of the Governor's budget at 1:00 A.M. Tuesday, passage of the sales tax hike at 3:11 A.M. Wednesday, and after four hours of debate commencing at 4:20 A.M. Thursday, passage of the Quality Education Act at 9:15 A.M. Not one Republican cast a vote for the QEA nor for any of the Governor's tax bills, although there were a few abstentions.

Although some Democrats disagreed with the Florio program—or saw it as being "too much, too soon," the party faithful fell into line with their leadership and the Governor, providing the bare minimum, simple majority necessary for passage of each bill: 21 Democrats in the Senate; 41 in the Assembly. A few Democrats, facing stiff Republican challenges in their districts, or strong opposition to the Governor's program from constituents, were able to abstain or vote "nay" with the party's blessing without endangering the Democratic sweep. On July 3 at a special ceremony, Governor Jim Florio signed the QEA into law.

Democrats expected the electorate to question their decisions, but hoped there would be sufficient time and opportunity for full explanation to voters back home before the 1991 election. Democrats hoped to demonstrate that over time, the homestead rebates, increased school equity, and no income tax increases for most New Jersey residents (the income tax bill had been revised to apply to single filers who made over $35,000 annually and joint filers and single heads of households who earned over $70,000 per year) would prove to be positive outcomes of the Florio initiative. Unfortunately, the strength of organized opposition would dash these hopes.

OPPOSITION, AMENDMENT, AND DEFEAT

A poll conducted by Newark's newspaper, *The Star Ledger*, and the Eagleton Institute of Politics at Rutgers, New Jersey's state university, between the July 2 and 10, showed that only 23 percent of the state's residents gave the Governor a positive rating and 68 percent gave him a "poor." This drop represented a significant change from the previous March when his election "mandate" was considered still effective. The drop was attributed by pollsters to "tax backlash" (Kerr, 1990). Hence, the Governor's popular support had dropped significantly because of the QEA and tax increases. Much of the loss can be traced to organized opposition from the New Jersey Education Association, two antitax groups, and their unusual alliance of convenience with the National Rifle Association. Each group had its own reasons for opposing the Governor's programs, illustrating the high political risk of trying to do "too much, too soon"—controversial multiple tax and wealth redistribution legislated in a single session. What did each group fight for and what were the effects?

Teacher Opposition

After Assembly passage of the QEA, on Thursday morning, June 21, large groups of teachers opposing the pension clause assembled outside the statehouse and met departing Democratic legislators with a chorus of boos and threats to vote them out as soon as possible. The NJEA had done an effective job convincing teachers that their pensions were in jeopardy, although the Florio administration believed state guarantees of local pension funding retained security.

The Antitax Opposition

In addition to opposition from teachers, two antitax organizations, United Taxpayers of New Jersey, formed in 1977 to protest the newly created income tax, and a new organization, Hands Across New Jersey, joined the opposition.

The latter, formed by several concerned citizens who met through response to a radio talk show on tax hikes, had 2,000 volunteers, coordinators in all the state's 21 counties, and 800,000 signatures by November 1990 supporting a repeal of the Florio tax package. The groups labeled the Governor as one who had gone back on an election promise not to increase taxes, even though no such promise had been made. In a 1996 interview, John Budzash, one of the "Hands..." founders, stated that the Governor had not offered adequate public explanation for the tax legislation and did not educate the public about the national recession, nor the spending policies of the previous administration, all of which led to the $600 million budget deficit. Budzash stated that Governor Florio should have emphasized that the small sales tax increase would probably have meant no more than 30 additional dollars per year to the average citizen (Budzash telephone interview, 1996).

The sales tax generated the most public reaction for the short term and took effect on July 1, the start of the new fiscal year when a new balanced budget was required by law. It hit all groups immediately, including the poor, and provided short-term evidence, apparent to the entire state, for the opposition's campaign.

The income tax increase, by contrast, was less an immediate issue since it would not take effect until 1991 and applied only to higher incomes. It would, therefore, be less obvious than the sales tax.

The property tax rebate plan did not quiet the opposition. Increased property tax rebates for low- and moderate-income residents would not be evident until 1991. Rebates are normally mailed just prior to elections, however, and the Democrats were depending on their help in November 1991 when the entire Legislature came due for reelection. The Governor had planned to suspend 1991 rebates to conserve funds but the Legislature had reinstated them in order to induce votes.

Opposition from the NRA

A third opposition group, an unexpected source of vocal and financial support for tax and teacher pension restructuring opponents, was the National Rifle Association (NRA). On May 17, 1990, the Legislature had passed an administration-sponsored bill restricting semiautomatic "assault firearms." It gave New Jersey the strongest law of this type in the nation and along with California, was the only other state to ban such weapons. From this point, the NRA and a local organization, The Coalition of New Jersey Sportsmen, became active lobbyists opposing the Governor. In a September 1995 newspaper interview, Joseph Doria, today Assembly Democratic leader, but Assembly speaker in 1990, felt that the income tax bill, passed a month after the weapons ban, gave the NRA a "linkage" to punish the Democrats. "The NRA couldn't turn us out on the assault weapons," stated Doria. "But they could get us on the tax. It was all NRA money behind the anti-tax movement" (Braun, 1995). Doria felt that weapons ban legislation should have waited until the following year in order to avoid linkage. Scheduling weapons legislation for May, Doria felt, was part of the "too much, too soon" problem. The NRA, however, remains unsuccessful today in winning a repeal of the ban with the new administration of Governor Whitman, herself a sports rifle enthusiast.

Media Opposition

Opposition forces received publicity from New Jersey's only statewide radio station, 101.5 F.M., which became a "talk" station to capitalize on the growing interest in Trenton (New Jersey's state capital) politics. The *Trentonian*, a tabloid newspaper, also exploited the opposition to the Governor and the Democrats with unflattering photos and reporting.

The Election of November 1990

The New Jersey election of November 1990 was the first indication that Democrats were in reelection trouble. The election included a Congressional contest between popular Democratic incumbant, Senator Bill Bradley, and a relatively unknown Republican challenger, Christine Todd Whitman. There were also two off-year state legislative contests, one in the southeast New Jersey 1st District, a region of split Republican and Democratic loyalties, and the other in the normally Democratic strong-hold 35th District of Paterson and Passaic counties.

The election was a shock to reality for Democrats. Senator Bradley was nearly upset by Whitman who captured 49 percent of the vote. Voters had no personal disagreement with Bradley's two-term U.S. Senate performance, but were simply sending a message that they opposed the Democrats because of the tax program of the Democratic Governor. In his campaign, Bradley had avoided the issue claiming it was a state concern and one that should not involve a Congressional contest. Bradley had also received Florio's support in the 1978 primary, one of only three endorsements from elected officials, and he did not want to critize the Govenor's program in 1990.

At the state level, popular Democratic Assemblyman Edward H. Salmon, a teacher, coach, and school administrator, was defeated by a Republican for the Senate seat vacated by a resignation earlier in the year. Salmon, who had been a legislator since 1987, approved the Governor's tax package but abstained from the school aid vote to avoid antagonizing fellow teachers over the pension issue (Salmon interview, 1995). Salmon had heavy financial backing from the state Democratic Party, including funds for videotaped messages for distribution to each voter. He lost nearly 56 percent of the vote. The defeat represented not only tax backlash but NRA-inspired opposition to the assault weapons ban from the numerous sportsmen and hunters in the rural 1st District region.

In Paterson, Assemblyman Cyril R. Yannerelli, who had been appointed the previous May to fill a vacated seat, was defeated by 2,840 votes out of 13,437 cast in a light registration and voting turnout. Yannarelli's vote for the sales tax increase had been crucial although he had upset Democratic leaders by abstaining on the income tax bill. He voted for the sales tax and school aid after party pressure. During the election, Yannarelli had been under indictment for alleged tampering with voter registration records. Although he denied the charges, the indictment hurt his election chances. But his defeat was interpreted as additional evidence of Democratic voter backlash against a Florio supporter.

Amending the QEA

Following the November election, Governor Florio announced formation of a blue ribbon commission to review the Quality Education Act for possible amendment with input from the public. Legislature Democrats, all of whom faced reelection in November 1991, began planning massive additional property tax relief. The QEA was amended in March 1991. Changes included a school aid reduction of $360 million with the money earmarked instead for property tax relief. This cut made sufficient foundation aid for middle-income districts impossible. The state also agreed to continue payment of teacher pensions and social security costs for another two years thus further reducing funds for school aid. After passage of the amended QEA, now called the Quality Education Act of 1991, or QEA II, the disparity in spending between wealthy districts and poorer districts widened. Education Law Center advocates, who had recommended a slower, "phased in" strategy for school equity funding as more politically feasible, now filed a motion with the state supreme court alleging that the reduction of school aid effectively gutted the *Abbott* mandated reform (Block telephone interview, 1996).

The Election Defeats of November 1991 and 1993

QEA amendments did not salvage the Democratic Party in 1991. Republicans swept both houses gaining a 58 to 22 edge in the Assembly and a 27 to 13 margin in the Senate. It was not quite a preview of 1993. In the interim, between 1991 and the fall election of 1993, education aid and teacher pensions were largely removed as issues through continuous dialogue between the Florio administration and NJEA. But the tax issue remained. Challenger Christine Whitman campaigned on a platform that promised a 30 percent reduction in income tax, the prime support for the *Quality Education Act*. The strategy worked, but just barely. Out of 2.3 million ballots cast, Whitman achieved victory by 30,000 votes.

The Democratic effort to achieve education equity through legislation had failed. But in the mind of James Florio, even today, it was a short-run defeat, a necessary risk of "too much, too soon" for the achievement of a long-term goal (Governor Florio interview, 1995). He may be correct. Time will tell.

EVALUATION AND CONCLUSIONS

The quest for education equity in New Jersey, as defined by the court in *Abbott* v. *Burke*, is not over. It will probably continue as long as there are public schools. The efforts of the Florio administration must therefore be viewed not only within the short span of 1990 to 1993, but in the indefinite long term, to gain an accurate assessment of their significance.

This political history has explored the course of Governor Florio's education equity plan with its related tax proposals. It has identified the issues and arguments of "too much, too soon," which apparently brought an eventual weakening of equity legislation through amendment, the 1991 defeat of the Democratic majority, and the 1993 defeat of the Governor by a slim margin. These issues, which compose the story of the 1990 to 1993 "short term," are summarized below. They are followed by a judgment of more long-term significance to real school equity.

The "Too Much, Too Soon" Short-Term Issues

If the record of the Florio administration is evaluated exclusively within the 1990 to 1993 time period, the election defeats of 1990, 1991, and 1993 stand out as the state's final judgment on the redistributive method of funding education equity. The defeats seem to indicate that too much controversial legislation was attempted without adequate time for review and the development of popular support through public education. This "too much, too soon" evaluation includes the following issues:

1. Mandate? Was the Governor's election margin a mandate for action making the first six months of the Florio administration the best window of opportunity for legislative action? Or was the mandate belief only a myth, and the "landslide" victory only an indication of the opponent's campaign mistakes?

2. Recession? The growing nationwide recession in 1990 cast doubt on the political success of any legislative program involving tax increases. Given this uncertainity, should the Florio administration have delayed their attempt to achieve education equity through satisfaction of the *Abbott* requirements until a more economically appropriate time? Or did the issue of equity justify pursuing costly school aid legislation in spite of the harsh economy?

3. Credibility? The recession produced a credibility issue for the Governor through a chain of events unforeseen during his 1989 fall election campaign. By January 1990 the slowing economy had reduced state revenue significantly below expectations and would produce a severe budget deficit by the spring.

Governor Florio proposed a sales tax increase to close the gap by the new fiscal year beginning July 1, and proposed, as well, an upper-level income tax rise to fill the deficit produced need for added school aid revenue required by the *Abbott* v. *Burke* school equity mandate. During the campaign, the Governor had stated that tax increases appeared unnecessary in view of outgoing Governor Kean's reports of an expected revenue surplus. Florio's statement, however, was publically construed as a "no tax increase" promise and when tax increases became necessary, many voters chose to believe the Governor had been deceptive. This perception caused a significant loss of voter support from which Florio and the Democratic Party were unable to recover.

Thus, the following issues are raised: Could the Governor have defeated the idea that he had gone back on an election promise if he had been a better "salesman"? Should Governor Florio's ultimate 1993 election defeat be attributed to lack of sales technique? Or was there simply too much to sell in too short a time period regardless of the tax issues?

4. *Tax Package?* Legislative leaders persuaded the Governor to abandon his original plan for QEA and income tax consideration later in 1990, and reschedule these proposals as part of a package with the sales tax in June. Would Governor Florio's original strategy have denied antitax interests their early strength and would it have given the administration time to successfully explain to the electorate the purpose and need for the full program?

5. *Assault Weapons Ban?* The assault weapons ban received Legislature approval in May, approximately one month before passage of the QEA and tax increases. Should the Governor have waited a year before seeking the assault weapons ban, thereby avoiding alienation of the NRA and their consequent funding assistance to the antitax and NJEA opposition groups? Did NRA support substantially broaden opposition influence on the voter?

6. *Exclusive Legislation Draft Process?* This issue relates to the Florio administration's planning of legislation for structural change within an environment exclusive of differing opinions, the strategy addressed by David Plank in his article: "Why School Reform Doesn't Change Schools: Political and Organizational Perspectives" (Plank, 1988). Would a broader-based plan, which allowed more time for compromise, have defused most opposition? If so, would broad-based planning have been at the expense of needed structural change, or would necessary change have been preserved with sufficient time for analysis of potential long-term impacts?

7. *Teacher Pensions?* The New Jersey Education Association (NJEA) believed that the decision to pass responsibility for teacher pensions to local school districts was made without sufficient time given for analysis of the long-term impact on pension security, salaries, and dollars available for the classroom (Corona interview, 1995). If teacher pension restructuring had been deleted from the QEA, in spite of its key school aid revenue purpose, would NJEA and teacher support have been assured? Would the full support of

educators have produced the decisive margin for Democratic victory in 1990, 1991, and 1993, reducing or eliminating the scapegoat role assumed by Democrats in those years?

8. Redistributive School Aid Financing? Governor Florio proposed taking income from affluent districts after a transition period, passing it to middle-income and poorer districts, and giving little or nothing tangible in return to the donor region, thus placing the full burden of school system financing on their local resources. In addition, some of the poor urban districts were suspected of corruption or inability to efficiently manage large amounts of aid revenue. Can wealthier areas, such as Bergen County, be expected to support such a school aid program? Were QEA amendments, which provided additional categorical aid and transportation aid to wealthier districts, a sufficient answer to their concerns about the redistributive approach?

9. The Administration's Style and Tactics? The motives of the Governor in pushing forward such a heavy legislative agenda within the brief first six months of his administration were questioned. Many legislators complained of having too short a time to review such a complex legislative package. Did these factors erode relations between the Governor and the Democratic legislators and in turn contribute to the Democratic losses in the 1991 election?

Success in the Long Term?

The foregoing suggests that the ultimate weakening of the Quality Education Act of 1990 and losses by the Democratic Legislature and Governor were caused by factors within the realm of political strategy, salesmanship, the state of the economy, and related issues of legislation overload and time. But the political defeats of 1991 and 1993 have a more substantive cause than the immediate circumstances of 1990. Florio's strategy of risk exposed essential problems born within the long-term history of the state with its diverse heritage and structure. The Governor's program was meant to address these more fundamental issues in order to create a long-term remedy, even though there existed the strong possibility of early losses. If and when these issues are addressed, the former Florio administration may yet see a long-term "post humous" victory.

New Jersey Heritage and Tradition

Statewide equal education opportunity, especially when defined in terms of dollar expense per pupil, was too radical for New Jersey in 1990 regardless of political strategy, the state of the economy, or effectiveness in "selling" the idea to the public. The state's tradition of local autonomy, a heritage brought by early Dutch and New England settlers, has created a multitude of 620 school districts, 78 in the relatively small geography of Bergen County, that generally resist not only consolidation, but efforts to share program expenses (e.g., advanced placement and special education courses, infrastructure). Consequently, the QEA idea of redistributing wealth from more affluent areas to middle-income and poorer urban and rural school districts, has no precedent and runs against state tradition.

Expressing a similar idea, Beth Bader, of the Consortium for Policy Research in Education at Rutgers University, suggests that public reaction to *Abbott* v. *Burke* did not consider education for children in poor districts a public good, and viewed education for children in wealthy suburban districts as a "private good," a good which the Reagan administration saw as beneficial to the person who had it but of little or no benefit to the viability of society (Bader, 1991, p. 7).

The fragmentation of school districts and municipal government in New Jersey also reduces opportunity for urban minority interaction with the wealthier suburban/rural majority, thereby reducing chances for overcoming resistence to a more regional or areawide outlook. In New Jersey, this condition not only has overtones of racism but suggests a suspicion of "third" world societies. The financially stressed Jersey City Public Schools, the inspiration for *Robinson* v. *Cahill*, must accommodate over 30 languages and the multitude of cultures this represents among its student population. Yet few incentives exist that would encourage the independent districts of suburban Morris County to share resources with Jersey City.

Bader also addresses the issue of racial overtones in school finance inequality, writing that it is a consequence of affluent whites fleeing to the suburbs, leaving the cities with a shrinking wealth base and an African-American/Hispanic student majority (Bader, 1991, p. 12).

Lack of state cohesion is also manifest in New Jersey's location between New York City and Philadelphia. Two distinct proprietorships existed in Colonial New Jersey, one called West Jersey which had a Philadelphia orientation, and the other East Jersey which focused on New York City. New Jersey still retains that split focus. The Philadelphia-oriented New Jersey of the Delaware Valley has relatively little interest in the New York City-oriented New Jersey of the Hudson Valley. The Greater Trenton Symphony, for example, has many key players from the Philadelphia Orchestra. Commenting on this phenomenon, Bader writes that while most major cities have independent resources, New Jersey's cities are largely satellites of New York and Philadelphia (Bader, 1991, p. 21).

Governor Florio's Risk for Long-Term Success

Governor Florio began a process to overcome this fragmented heritage by shifting the political paradigm from one of distributive politics, to use Theodore Lowi's term, in which all key actor groups would receive more resources under state aid, to redistributive politics in which the poor are subsidized at the expense of the rich. Through the political risk of this legislated redistributive education equity, presented in a concentrated format within a brief time frame, the Governor brought New Jersey face to face with itself and gambled that the population would understand the long-term need for a more common statewide identity.

But during the short term of 1990 to 1993, the politics of consensus broke down even to the point where urban and middle-income suburban districts that were supposed to benefit from the *Quality Education Act,* failed to realize their new found advantage as their attention was diverted by the tax issue and the statements, some misguided, of the opposition coalition.

The program envisioned by the Governor and the plaintiffs of *Abbott* v. *Burke* can only succeed if the state develops cohesion, a greater interaction among its urban, suburban, and rural communities, school districts, and regions. This conclusion is supported by Jaffe and Kersch who argue that school equity has wide-ranging, extra-legal dimensions. Disparities must find remedy in a solution "generated and embraced" by a broad constituency (Jaffe & Kersch, 1991, p. 297).

The issues raised by the Florio administration forced a debate of this potential, an often-heated dialogue that continues today. The Governor purposely designed a legislative program, that some would call "too much, too soon," so the state would face the political, social, and economic realities of the equity problem. He took a political risk initiating a process that otherwise might not have occurred and thus created a legislated solution to the equity problem. Should the long-term result in a measure of equality among New Jersey's public schools, some credit must be given to Governor Florio and his "push at the edges" philosophy which advocates "determining what is politically possible and then doing more" in order to bring essential structural change—"to really excel" (Budget Address of Governor James Florio, March 15, 1990).

Other states in quest of school equity through legislation will face their own historical, political, and social conditions, perhaps quite different from those in New Jersey. But these states will undoubtedly meet similar unknowns when considering what is "too much" or "too soon," the same risks of short-term defeat when attempting more than the politically possible in order to "excel." Education equity brings a burden of choice to elected state officials, be they executive or legislative, a choice between the politically safe course which avoids the equity question, or the path that "pushes the edges," the risk that may bring election defeat, but equality of opportunity for the long term.

INTERVIEWS

Steve Block, The Education Law Center, Newark, NJ (telephone interview) 8/6/96.
John Budzash, "Hands Across New Jersey" Anti-tax organization (telephone interview) 6/4/96.
Dolores Corona, Director of Government Affairs, New Jersey Education Association. 7/12/95.

Kathleen Crotty, Senate Democratic Minority Office Director, State House, Trenton, NJ. 9/15/95.

Joseph V. Doria, Assembly Democratic Minority Leader (31st District, Bayonne, Hudson County), telephone interview, 6/7/96; personel interview 6/21/96.

Hon. James J. Florio, Esq., Florio & Perrucci, Piscataway, NJ. 12/26/95.

Paul Muller; Jason Teele, Assembly Democratic Office, State House, Trenton, NJ. 6/27/95.

Edward Salmon, NJ Commissoner of Public Utilities, Trenton NJ. 12/18/95.

Nathan Scovronick, Project for New Jersey, Woodrow Wilson School, Princeton University, Princeton, NJ (telephone interview) 5/21/96.

REFERENCES

Bader, B.D. (1991). *Abbott v. Burke v. New Jersey: Policy, politics and political economy.* Consortium for Policy Research in Education, Rutgers University. Paper presented at the AERA Meeting, April 6.

Braun, R.J. (1995, Sept. 15). Doria picks over his meal and Democratic bones. *The Star Ledger*, p. 27.

Corcoran, T., & Scovronick, N. (1994, unpublished). *More than equal: New Jersey's Quality Education Act.*

Florio, Governor James. J. (1990, March, 15). *Fiscal year 1991 budget address.* Trenton, NJ.

Goertz, M.E. (1992, Nov). *The development and implementation of the Quality Education Act of 1990.* New Brunswick, NJ: Consortium for Policy Research in Education, Rutgers University.

The Governor's Education Policy Task Force (1994, April 7). *A school funding proposal beginning 1995-96.* Trenton, NJ.

Jaffe, M., & Kersch, K. (1991, Summer). Guaranteeing a state right to a quality education: The judicial-political dialogue in New Jersey. *Journal of Law and Education, 20* (3), 271-300.

Kerr, P. (1990, July 15). Florio popularity crashes in survey. Drop in ratings since March is tied to tax increases. *The New York Times*, p. L 19.

Lehne, R. (1978). *The quest for justice: The politics of school finance reform.* New York: Longman, Inc.

Lowi, T.J. (1972, July/August). Four systems of policy, policy, and choice. *Public Administration Review*, 298-310.

Plank, D.N. (1988). Why school reform doesn't change schools: Political and organizational perspectives. In W.L. Boyd & C.T. Kerchner (Eds.), *The politics of excellence and choice in education* (pp. 143-152). New York: The Falmer Press.

Chapter 3

COST EFFECTIVENESS? OR EFFECTIVENESS AT A COST? EAI AND THE HARTFORD, CONNECTICUT PUBLIC SCHOOLS

Thomas H. Jones

In October 1994 the Hartford, Connecticut school board hired Education Alternatives Incorporated (EAI), a private company, to operate its entire 26,000 pupil, K-12 public school system. The company was fired in January 1996 ("EAI The Hartford Experiment," 1996). The Hartford contract marked a breakthrough in at least two respects. It increased the company's operating revenues many fold. From a public policy viewpoint, Hartford is the only school district in the nation to have ostensibly turned over its entire school system to a private operator.

Advances in Educational Productivity, Volume 6, pages 43-65.

ISBN: 0-7623-0082-5

This paper begins with a background section on EAI. The corporation's origins, leadership, educational programs, and financing are examined. The paper continues with a more detailed examination of EAI's activities in Hartford. What did the city's school board expect? The contract is analyzed; events leading to its breakdown are chronicled. The short-term impact on Hartford's public schools is assessed.

As of this writing the EAI experiment in Hartford is over. Nevertheless, profit-seeking corporations continue to offer educational services in a variety of ways. This development holds profound implications for school administration and finance. For these reasons a second purpose of this paper is interpretation of the school reform efforts in Hartford.

An analytical section explains the inherent conflict between EAI and the Hartford School Board. In the final section of the paper, I draw some more personal observations about the implications of events in Hartford, viewing the EAI experience as a case study with wider applications.

BACKGROUND

Educational Alternatives Incorporated is the creation of its 51-year-old founder and chairman, John Golle (pronounced "Goalie"). Like his better-known competitor Chris Whittle, Golle's background is not in education, but in technology and in sales.

The son of a moderately successful businessman, Golle began his career 28 years ago as a copier machine salesman for the Xerox Corporation. His sales successes with Xerox led him to start his own sales training company, Golle & Holmes. In that role he trained the sales forces of some of America's largest corporations, becoming acquainted with many CEOs in the process. Golle & Homles was located in Minneapolis, a city known for its progressive and socially responsible business climate. It is the Twin Cities that Golle knows best, and where his corporate connections are the strongest.

By the mid-1980s Golle had reached mid-life and amassed a multimillion dollar fortune. Feeling too young to retire and wanting to do something more meaningful with the rest of his life, the education field seemed attractive to him.

In 1986 Golle bought the business plan for another Minnesota corporation, Private Schools Development Corp., a predecessor entity to the Tesseract Development Corporation.[1] The Tesseract Model forms the basis for Golle's educational view, later implemented through EAI. It is to that model which we now must turn.

The Tesseract Educational Model: Schooling "To Go"

The Tesseract story begins with a Minneapolis business acquaintance of Golle's, William Norris. Norris, chairman of the multibillion dollar Control

Data Corporation, believed that schooling was excessively labor intensive and antediluvian in its methods. It employed essentially the same teaching and learning techniques as those prevailing in the last century: chalkboards, lecture, paper and pencil testing. Corporate America could improve education and make money too.

To that end Norris founded Private Schools Development Corp. as his sideline business. But when Norris's major business, Control Data, foundered, he wanted out. Golle bought the company.

As the corporate name suggested, the market identified was in the nonpublic school sector. There was, Norris believed, a vast number of middle-class parents who wanted quality education for their children not available in most public schools. These parents could not afford the high costs and amenities associated with the elite independent schools. On the other hand, these same parents were looking beyond the low-cost, home-grown parochial academies for a "state of the art" education including science, math, and technology.

Norris himself never opened any schools. Private Schools Development Corp. was really nothing more than a business plan. To implement the plan Golle involved another Minneapolis enterprise, the mid-size computer software firm: The Ceridian Corporation. (Ceridian was a corporate successor to the defunct Control Data.) Ceridian, in turn, had a wholly-owned subsidiary called Tesseract—an entity that operated a Minneapolis school aimed essentially as a proving ground for software. Forming EAI in 1987, Golle bought operating rights to the Minneapolis school building and a 50 percent share in Tesseract. By the late 1980s Golle was operating a second Tesseract facility in Paradise Valley, Arizona. His role as an educational entrepreneur was launched.

Golle, of course, bought into Norris's original vision. He reoriented the two EAI schools from software development to education itself. Central to that vision is the computer connection. Emphasis was placed on high levels of pupil achievement combined with strict cost controls. Computer-based learning was core for most school subjects. There were personalized education programs for all students, including special education students. Reasonable class sizes were to be achieved through minimal administrative costs and differentiated staffing. Staff in-service training would emphasize computer software use. Parents would be involved.

In the late 1980s cost overruns forced EAI to raise tuition from an initial $3,500 to nearly $5,000. Schools lost enrollment as a result. But Golle was aware that, even at the higher price, he could provide school services at lower cost than many public schools. By 1990 Golle decided to reorient his efforts once again, this time targeting the much larger public school market. The slightly amended new vision concentrated not on developing one facility at a time, but on putting together a comprehensive package of educational services for sale to the nation's public schools.

The package contained both instructional and instructional support elements. The Computer Curriculum Corporation would design proprietary classroom software for teaching and learning. Johnson Controls, still another large Minneapolis corporation, was to be hired as manager of buildings and grounds. And no less an entity than KMPG Peat Marwick, arguably the nation's leading firm of financial consultants, signed on to develop appropriate accounting and financial control systems.[2]

The service package could be offered to many public school districts simultaneously. If most of them said "no" there were few fixed costs—only a sales call. On the other hand, the volume of services could expand rapidly as cities and school districts signed on. Like MacDonald's hamburgers, EAI could keep its educational services "in the freezer" until customers arrive at the drive-through, then prepare them quickly, on demand. Though not a household name like MacDonald's, EAI had allied itself with some of largest and most respected corporate entities in the nation.

"Breaking the Mold": The Miami and Baltimore Contracts

Since the focus of this paper is Hartford, Connecticut, EAI's earlier public school operations will be mentioned only briefly, insofar as they may have effected Hartford events.

The most basic and compelling point that can be made about all of EAI's initial efforts is this: the only school boards that would hire EAI are those in which the traditional methods were perceived as failing. Mayor Schmoke of Baltimore summed it up: "Golle, I don't know if you're as good as some people tell me you are, or not. But you can't do any worse" (American Federation of Teachers, 1994, p. 2). The company has taken on some of the most difficult school situations in the nation. There have been no easy battles.

In 1990 EAI got its first public school management contract, South Pointe Elementary School, Dade County, Florida. The contract was drawn on a "fee for consulting" basis, and, arguably, represents EAI's most successful public school management effort to date. The contract was for five years, largely funded by private donations. Though results are ambiguous, pupil attendance improved as did test scores.

Ambiguity results from the fact that similar improvements occurred in a "control" school matched to the EAI experiment. Though the Dade County School District labeled the experiment "successful," the contract was not renewed (Stecklow, 1995). There were some disagreements over final payments, but the 1995 termination was reasonably amicable.

EAI's second public school contract in Baltimore, Maryland began in 1992. There were great hopes, and national press attention. On EAI's "first day of school" in Baltimore, Secretary of Education Lamar Alexander gave the effort a national award.

EAI was hired to run nine schools. The rest would remain within the traditional city system, inviting a sort of informal experimental group/control group comparison. Statistical comparisons were offered as justifications for a priori views. If EAI pupils achieved better scores on a particular test, that merely verified the existence of a temporary "Hawthorne Effect." If EAI pupils did about the same, or worse than pupils in the "control" schools, this proved the emptiness of the Tesseract model and its lack of any compelling rationale.

Other, diametrically opposite interpretations were possible. High EAI pupil scores on some tests demonstrated the experiment's obvious benefits. If EAI scores were lower, there obviously had not been a long enough period of time for EAI to show the success of its methods.

In purely financial terms matters were just as ambiguous. The 1992 contract called for the company to receive the same per pupil allocation as other Baltimore schools, about $27 million in the aggregate for that year. EAI had to abide by the teachers' union contracts; otherwise they had broad latitude in spending the money as they saw fit. An initial move was to spend about $1 million on improvements to the physical plant, then attempt to lay off school custodians, or reassign them to the corporate contractor, Johnson Controls. Claiming a need for further cost savings, problems emerged with the retention and pay of teachers aids. There was a high turnover of employees in this category. The teachers' union moved from mildly hostile to openly hostile.

EAI, arguably, wrongly counted some corporate expenses as school expenses. Likewise, city schools have extra expenses EAI doesn't. (Running a central office and out-of-district special education placements are examples of the latter.) Elaborate methods of cost attribution were devised, but they didn't resolve anything. If a particular calculation showed EAI spending a little less per pupil, certain corporate costs must have been omitted. If EAI spent more, this could be justified because of the physical improvements to EAI schools, and so on.

In May of 1995 the Baltimore City Council voted to subpoena the corporation's financial records. Mayor Schmoke had ordered a more general performance review of EAI earlier in 1995, but this open rift with the council proved the beginning of the end for EAI in Baltimore.

The anecdotal evidence on pupil and parental satisfactions is a little less ambiguous. Most observers liked the cleaner and newer facilities associated with EAI schools. However the Baltimore contract was terminated in late 1995, owing not so much to EAI's clear inadequacies as to Baltimore's financial problems. In the political matrix of that declining city, an outside contractor like EAI simply lacked the necessary clout. The winner of Lamar Alexandar's "break the mold" award had itself been broken by the mold.

Though relations with the Dade County/Baltimore school system were rocky, for the most part Wall Street liked what it saw. It is to that arena which we now must turn.

Wall Street and EAI: Caveat Investor

The vision of William Norris has become, by the mid-1990s, almost a banal observation: there is a wide market for alternatives to the local public school based on neighborhood assignment. In the political realm this manifests itself in the promises and actions of elected leaders. Discussion and legislation abounds concerning charter schools, magnet schools, vouchers, and tax credits.

In the world of corporate finance the same feelings are manifested in the prices of "education industry" common stocks and the hoopla surrounding every development in the industry. There is now an education industry index (up 54% in the first nine months of 1995). It tracks the performance of stocks in seven subsectors: (1) preschool, (2) general management, (3) product, (4) service, (5) at-risk youth, (6) post-secondary, and (7) vocational schools.[3]

Very few companies in the education industry can be counted on to make a yearly profit. Like most others, EAI has offered its long-term investors a rough ride. At various times in its brief existence its price per share has fluctuated between the low single digits and the high forties, trading mainly on expectations about gaining another contract or loosing one presently held. EAI has logged a small profit in only one year of its existence. However, that does not mean that there is not money to be made. Both the financial and political climate are ideal.

The current climate, plus the considerable talents of John Golle himself, go some distance to explaining the successes of the corporation to date. Golle understands that share prices, particularly in a trendy field like the education industry, are not good barometers of a fundamental outlook. Beneath the day-to-day fluctuations his activities suggest an orientation to the long term and a very considerable talent for corporate finance, though whether he will be successful ultimately is anyone's guess.

Considering Golle's long-term strategy, two elements stand out. First, Golle has used an "expansive" definition of company sales, revenues, and cash flow. Second, he has been willing to incur the wrath of unions by substituting physical capital for human capital wherever possible. What may seem politically dangerous is very nearly essential for EAI's corporate prospects.

With regard to the first point, Wall Street naturally likes companies enjoying rapid growth in their rate of cash income. In fact, cash flow is generally considered as a key long-term indicator of corporate value, especially in "start-up" companies.

Golle has taken the tack of booking *the entire school appropriation* as a "sale," even though the vast majority of the funds, perhaps 90 percent in urban public school systems, *must contractually be allocated* to the schools' present personnel. What might look to some as "the payroll account" EAI books as a "sale."

Dow Jones reports that EAI's gross sales have risen from $2.2 million in 1990 to $34.4 million in 1994, a calculated average annual growth rate of nearly 98 percent per year over the period.[4] On Wall Street this is the stuff of dreams! Of course, when a Baltimore or Miami contract ends, "sales" drop, and EAI's stock price takes a tumble. But the important long-term objective has still been served: the entire school budget is booked as corporate revenue. Sales growth of this sort is a much more enticing prospect to the investment community than the same firm reaping a few million dollars in consulting fees, though arguably educational consulting more nearly describes the firm's work to date.

The Tesseract Model has a second major financial implication. By "managing" the schools, physical capital owned or leased by EAI *but placed in the schools* can be fully depreciated as a business expense. Putting more physical capital in the schools—computers and the like—is arguably an educational advantage. But without question it is advantageous for corporate finance. One major financial service defines corporate value as a function of earnings plus depreciation.[5] In this construct, a dollar's worth of depreciation is as good as a dollar's worth of earnings.

Physical capital can be depreciated, but human capital cannot be. Thus it pays EAI, or any other private firm, to substitute the former for the latter where possible. The labor-intensive nature of public education, its lack of technological substitution for over a century, and the sheer size of the enterprise make schooling an attractive prize for financial "risk capital."

Golle's Dade County and Baltimore corporate experiences suggest the following lessons. Avoid claims of quick improvement in pupils' test scores. These did not work out in either of the first two efforts. Second, avoid wrangles with locally elected public bodies over EAI schools' expenditure accounts. Turning over the school budget to the corporation proved a difficult modus operandi, raising questions of cost comparison and specific expenditure decisions. At the same time, continue to offer EAI's services as those of a school manager rather than a school consultant. It is the former that would sustain favorable views of the corporation on Wall Street. All these lessons from the prior two experiences are reflected in the Hartford contract.

EAI AND THE HARTFORD SCHOOLS

Though a small city of only 136, 000 people, Hartford reflects, in microcosm, the worst problems of the nation's urban schools. The problems are inherently bound up with the residential segregation of races and economic classes. City schools are 93 percent "minority." Median family income is about 60 percent of the state average. The wealthy adjoining towns of Glastonbury and Wethersfield are less than 10 percent minority and have median income levels well above the state average.

Hartford, along with Connecticut's other school districts, is fiscally dependent. District boundaries are coterminous with the state's general units of local government called "towns." Colonial in their origins, Connecticut towns are small by national standards, both geographically and in terms of school enrollments. In the contemporary time period travel among the state's 169 towns has become fast and easy. No town in Connecticut is more than a two hour drive from any other town. These facts condition the environment in which Connecticut educational and social policy is made.

As a highly urbanized state, Connecticut is, in many ways, a single large metropolis with many local governments. Governmental fragmentation may have contributed to the substantial concentrations of wealth (and poverty) *within* towns, and substantial economic differences *among* them.

This situation is by no means unfamiliar to school finance specialists. A subtle difference in Connecticut is the small size of its jurisdictions, Connecticut had no annexation movements; its cities simply never grew in size beyond their eighteenth-century boundaries. As a result, a state known for its wealth nevertheless has some of the poorest and most troubled cities in the nation. National attention recently given to the City of Bridgeport's "bankruptcy" is only one case in point. Hartford ranks fourth and New Haven ranks seventh on one national poverty index. This does not imply, however, that all Connecticut cities are necessarily property poor. Some still have substantial amounts of commercial property.

A recent court case, *Sheff* v. *O'Neill*, centered Connecticut's attention on the problem. Regional, cooperative, voluntary plans are again under consideration. But progress is slow. Statewide, 80 percent of the state's minority student population continue to be clustered in 18 of 166 school districts while 140 others are more than 90 percent white.

The pupil testing movement has posed problems for policymakers and the public. If the criteria are set high, many city children fail and the state appears to be doing poorly. If the criteria are set low, more students pass the test. But there are claims that the test is unrealistically easy. Regardless of what test is constructed and where the bar is set, comparisons between Hartford and almost anywhere else are invidious. For example, in urban areas only 21 percent of eighth-grade math teachers even claim to emphasize algebra.

The 1993 *Excellence Commission Report* was a statewide effort to produce greater efficiency in schooling. It proposed providing greater use of school-site management techniques and greater reliance on public school enrollment options such as magnet schools (through Hartford's suburban busing program with only luke-warm city support). The report also raised the possibility of implementing the concept of "receivership" as a last resort. Under the receivership plan, the state would assume direct control of those districts and/ or schools that consistently do poorly on achievement tests. Finally, there are small but active school choice groups in Connecticut. (A modest charter school initiative succeeded just this year.)

Economically, the state relies heavily on the national defense and financial service industries. Educational enhancements were easier to finance in the 1980s, and that decade was very kind to primary-secondary education finance. Teachers' salaries moved from sixteenth to first in the nation. Connecticut achieved one of the nation's highest per pupil expenditure levels and lowest pupil-teacher ratios. In the 1980s the state's share of total public school spending rose from 32 to 45 percent, with most of the new money concentrated in the state's poorest districts.

The present decade has not been so kind. The recession, the end of the cold war, and layoffs in the financial services industry have meant slow growth at best. Many of the state's economic indicators show decline. A broad-based personal income tax—the first ever for Connecticut—was implemented in 1992. But it paid for past growth in school spending. Since the income tax, the state's share of total school spending has declined.

Corporate downsizing in the insurance industry, the closing of the city's major department store, and the move of some corporate headquarters out of state have hit Hartford particularly hard. Yet developments in the state and nation have meant that the city perceives little further help coming from external sources.

Distinctions between Hartford schools and its teaching staff mirror those described above. Most teachers are white in a city which is largely minority. Very few teachers live in the city. Still, the city remains one of the last bastions of strong unions. Hartford's teachers are among the best paid in the state and nation. Many classroom teachers salaries now exceed $50,000 a year. Per pupil expenditures exceed $10,000 a year, though the average elementary classroom is lucky to receive half this sum.[6] Salary accounts for administration and special pupil services are high. These were the conditions confronting John Golle as he approached the city's nine-member board in early 1994.

Happy Days: EAI and Hartford Sign A Contract

By the mid-1990s it became abundantly clear that substantial new money for educational innovations would not be forthcoming. The city would be hard-pressed just to remain in a steady state. Yet "school outcomes," defined mainly as boosting test scores, was an issue that would not go away. Second, there was a "conditions of learning" issue. Some school buildings were dirty and malfunctioning. There was arguably too much administration. The part-time, all-volunteer school board felt it was in a position to be dominated by the unions and professionals in the system. It knew it was somewhat out of control.[7]

The contract between EAI and the Hartford board, negotiated in the summer of 1994, can be viewed as a process of aligning Golle's aims and past experiences with the board's needs. Golle wanted control of the entire school system, not just a few schools. Comprehensive district management would immediately boost EAI's corporate "sales" many-fold; furthermore, management of the entire system would eliminate possible fights over disposition of school funds of the sort that had occurred in Baltimore. Full takeover would diffuse the test-score issue. Invidious comparisons between EAI schools and other schools in the same city would not be possible, since every school would be an EAI school.

Considering test scores more broadly, Golle offered no "quick fix," saying that any improvements in that area would take at least three years. It is perhaps surprising to some that the board dropped the issue so quickly. However, the board hoped that EAI's takeover of the management operations would free educators to concentrate on test scores, implementing a strategic plan that the board had set.

An American Federation of Teachers report harshly critical of EAI's test-score results in Baltimore may have actually redounded to the corporation's benefit.[8] In any event, both sides were persuaded by the reality of the situation: when it comes to test scores, there are no quick fixes.

This left the "conditions of learning" issue as the area where the board and Golle had a congruence of interests. Using Tesseract techniques, Golle could offer computer-based curricular innovation. Without raising costs for the city he could still make a profit for EAI. He could offer the promise of cleaner buildings. Reliance on corporate-type financial management suggested greater value would be received for the city's money. In sum, the contract can be viewed as the board's effort to reassert its control over recalcritant unions and professionals in the school bureaucracy.

The contract called on the corporation to provide:

> ...comprehensive educational and management operating services for the public school system... EAI shall continually assess, advise, recommend, and subject to the final authority of the Board, implement various enhancements to the educational program...to improve student performance...including, but not limited to, staff training, equipment, software, student evaluation...technological tools...TESSERACT...and other improvements.[9]

The board would continue to hold "...ultimate decision-making authority..." as prescribed by statutes.

Superintendent Eddie Davis, a career employee of the Hartford schools who had publicly advised against hiring EAI, would continue as chief executive officer of the school system. He, together with the board, would have complete control over matters involving all employees, "...hiring, assignment, duties, compensation, discipline or discharge." EAI was directed to "...work directly with (the superintendent) and subject to his authority...." If conflicts arose, "...at the request of either of them the issue shall be presented to the board jointly for a final decision...."

Below the board and superintendent levels, issues of power and authority were unspecified. EAI was committed to retain all current personnel and to follow all contractual agreements with the unions. The company would hire a local project manager and a business manager. Persons hired for those posts would be subject to board approval. Hartford's entire annual school appropriation, about $170 million, was to be "paid, controlled, and expended" by EAI. In the process, the company would develop its own management information and financial accounting systems to get a handle on where the money was going.

City payments to the company were to be made in 12 monthly installments. Compensation would be based on "any gross and net savings realized in any school year." Gross savings were defined as the annual school appropriation minus expenditures and encumbrances. To calculate "net savings" EAI's Hartford expenses were to be subtracted from "gross savings." If any portion of the annual school appropriation remained in the bank, it was to be divided on a 50-50 basis between the company and the board. Two-thirds of the board's share would be remitted to the city as a "gift." The remaining one-third of the board's share (1/6 of the total "savings") would be given back to EAI to be spent at the board's discretion (clause 4.5.2). Upon written notice either party could terminate the contract within 90 days.

To assure that the city did not respond to any savings realized by reducing the annual appropriation for the following year, the contract called for minimum annual increments in the school appropriation based on the consumer price index.

Over and above the annual school appropriation, EAI represented that it would expect to "invest" substantially in improvements to the school system's physical plant and equipment. The amount discussed was in excess of $20 million over the five-year contract's life. Most would be spent up-front, perhaps some $14 million in the first year. Items included school repairs, new photocopying machines, fax machines, snow blowers, and walkie-talkies for school communication in every school.

Hartford was happy. The promised capital investment items carried the contract to approval. The board's hopes for a more modern, computer-based learning system could be achieved without additional cost. Schools would be neat and clean, and the board itself would finally exert appropriate management and fiscal control.

John Golle was happy. Company "sales" would again increase sharply, as they had after the Baltimore signing. Capital depreciation would kick in as EAI acquired new school equipment through ownership or lease. The contract called for EAI to appoint a project manager and a business manager to run the Hartford project as Golle himself sought new clients.

Investors were delighted. EAI's stock price headed north in the second half of 1994. A new day in public education appeared to be dawning.

The "Sweetheart Deal" Sours: EAI Gets the Boot

Hartford's EAI privatization experiment lasted little more than a year, from October 1994 to January 1996. Board votes on matters effecting EAI always reflected deep divisions. EAI never held more than a five to four majority and could not command even that level of support on crucial decisions. A priori ideological views probably doomed this particular experiment from the beginning. But ideologies change. Tracing the history of events leading to the contract's cancellation may be useful to understanding future efforts.

In November and December 1994 EAI delivered substantial amounts of capital equipment to the schools and conducted some staff training in their use. However, a disagreement over computers led to the first serious public altercation. EAI received a special price on Dell computers while several schools wanted Apples. Golle weighed in on behalf of Dell, saying that both machines could do the same things and Dells were a better value. This was an example in microcosm of what is wrong with school decision making. "You're stiffing yourselves," Golle said.

In this first significant disagreement between the company and school personnel the board backed the schools. EAI acceded and agreed to provide either machines, at the discretion of the principal. New computer labs were installed in five schools throughout 1995.

In February 1995 plans were announced to reorganize schools and "improve worker productivity" by implementing more elements of the Tesseract model, but these corporate plans had to be postponed. It became clear that the sour economic climate of the state had finally overtaken Hartford schools. There was a $3.3 million deficit. Hartford teachers reluctantly accepted a wage freeze.

An early retirement plan for non-certified staff helped to cut the deficit. By all accounts, EAI's financial advice was crucial in closing the gap. Superintendent Davis acknowledged EAI's help. Several board members credited EAI with "saving" millions of dollars ("EAI the First Year," 1995). Resolution of the budget shortfall marked a high point in EAI-city relations. However, the matter proved costly to those relations in the longer term.

To "manage" a $165 million school budget and affect the reorganization of 32 schools and 3,000 (mostly hostile) employees, EAI never had more than 15 personnel on site in Hartford, and this number included employees of the subcontractors, Johnson Controls and Peat Marwick.

The small number of EAI staff in relation to the size of the entire school system operation severely limited the company's control. Mainly involved with long-range strategic planning, people on site simply became overwhelmed with the school system's day-to-day operations. EAI employees were charged with missing appointments, not returning phone calls and not participating in community events.

Finally, in March the company's acting business manager proposed that EAI take over all accounting operations, as the contract had required from the beginning. There were reasons for the delay; EAI wanted to avoid criticisms of its financial management of the sort that had built up in Baltimore. But a project manager still had not been hired; new accounting system software had not been installed—there were the inevitable wrangles over who would pay for it—; and the company's Hartford office remained grossly understaffed. Decisions of any consequence—for example, the Dell versus Apple computer flap—had to be referred to Golle himself, and he was in Minneapolis, or on the road seeking new clients. Despite the contract language, the board refused EAI's request to take over day-to-day operation of the accounts.

In April Superintendent Davis submitted a budget proposal for the following year. The board rejected it, telling Davis to work more closely with EAI. Davis took a "hands off" attitude, essentially letting EAI give the budget a try. The following month EAI presented its version of the annual budget to the board. The document reflected exactly the kind of changes that reasonably could have been expected: downsized school staff, more reliance on computer-based learning, and contracting out school support services. In composite the proposal meant a major reorientation.

Of course, EAI's budget proposal would have resulted in major downsizing. Savings in staff salaries would be directed to the corporation's subsidiaries and subcontractors. The board rejected EAI's budget proposals as self-serving, and indeed a budget much closer to Davis's original proposal was ultimately implemented.

By late spring relations between the company and the school administration had fully broken down. The company submitted to the superintendent a million dollars of vouchers for its own, company expenses. Davis refused to pay, claiming that EAI could be reimbursed only out of savings, and there were no savings. Requisitions submitted by building principals continued to be processed as usual, by board administrators. The small, overworked Hartford staff of EAI was still being bypassed.

Whatever mutual respect had been generated by EAI's February financial rescue package was destroyed by late spring. The board was weakened by long-standing structural incapacities to make independent analyses and by internal divisions over the merits of this experiment. Minimal though it was, board support was the only support EAI had. Canny investors slowly became aware that the corporation could not control the money it supposedly managed, even to pay its own expenses.

By early summer it became clear to corporate headquarters in Minneapolis that they were in a catch-22 position. The small staff assigned to the Hartford project was grossly inadequate for the sheer volume of work required. Yet having been refused payment for company expenses, Minneapolis was hardly willing to expand its Hartford operation.

The company decided to fall back to its Baltimore model. It proposed to "operate" only five schools during the 1995-1996 school year. The rest would be fully returned to system control by mutual agreement.

The new EAI proposal further divided the board. Board President, Thelma Dickerson, waxed enthusiastically about the downsized effort. This will give EAI a chance to "transform those five schools from top to bottom," she said. When school opened in September, however, EAI had not completed its physical plant improvements even in those five schools. Board member Carroll had a different view, "My confidence in them fell dramatically. I almost called for their ouster, but it would have been seen as a big victory for...the status quo." At the same time it was downsizing its actual operations, EAI was still insisting on booking the entire school budget as a company revenue.

The contract had not been abrogated by either party; it allocated the entire annual appropriation to the company. On the other hand, the contract also allocated operation of the entire school system to EAI. Could EAI claim financial control over schools it did not manage? This inconsistency drew sharp criticism from the city's financial officials.

The fairly arcane financial questions were suspended during the fall 1995 school board election campaign. Though there were three slates, there was only one issue: EAI versus "the unions." Most Hartford teachers live in the suburbs, are of a different race and social class than the students they teach, and they had difficulty organizing city voters. In a light turnout, the vote tilted slightly toward EAI. On the newly elected board, EAI retained its nominal majority of five to four. Davis authorized a token $350,000 payment to the corporation.

The school-board elections proved to be an anticlimactic distraction. Late in November EAI lost its Baltimore contract and reported a multimillion dollar loss for the year. Golle made it clear that most of the responsibility for the corporation's financial problems were due to the intransigence of Hartford officials. He promised stockholders a new corporate strategy, one emphasizing suburban school districts instead of the highly politicized cities. In January 1996 three of the five-member EAI majority on the newly elected board called a press conference and announced they would no longer support the experiment. This event effectively ended EAI's formal role.[10]

A host of legal and educational questions remained. How would EAI's claims to property located in the schools be addressed?; Was EAI entitled, under the contract, to reimbursement for any of its expenses?; and, What would the city and school board do to improve education in Hartford? By late 1996 these matters had been settled to the satisfaction of both parties.

ANALYSIS

Webster's Dictionary defines "management" as the "direction of an enterprise." The synonym is "control." By this ordinary definition, EAI never had been the manager of Hartford schools. It never had the power to direct any teaching and learning, to assign any student, to hire or fire any school employee, or even to write a job description for implementation in the future. Though its contract with the board gave EAI control over school funds, the corporation was never permitted to exercise that control. The legal, formal organizational structure of the school system never changed. All decisions remained with the superintendent and the Hartford Board of Education, just as they always had.

Part of the reason for no change can be assigned to laws and statutes. Connecticut statutes are like those of most states in that they vest major powers and responsibilities in local boards of education. These responsibilities cannot legally be abrogated or assigned. Nevertheless, the board did have wide powers to accept EAI's advice and almost consistently refused to do so. In smaller matters, and bigger ones, the board sided with its own employees against EAI.

Conditions surrounding the "sale" of EAI's services to the Hartford board were full of half-promises and misconceptions. Some of the confusion may have been self-serving, for both sides. Local news reports emphasized the new computers and other equipment, not the layoffs that would shortly have to occur to pay for these "corporate investments." There was a public sense that EAI would do just about anything to get the Hartford job. In advertising parlance Hartford would be a sort of "loss leader" for the corporation. By using Hartford as a proving ground and showing how well they could do, other cities would flock to buy EAI's services. The corporation would gain enough new business to make up for any Hartford loss. That is the reason that the corporation was willing to offer their services on such advantageous terms: they would be paid only out of "savings." The phrase, "loosing on this sale, but making it up on volume," has become a old joke. The Hartford Board should have known better.

For its part EAI knew exactly how it planned to enrich corporate prospects through the Hartford experience, but they never said clearly and publicly how this was to occur. They talked about computers but not about capital substitution—physical for human. They talked about efficiency, but the contract language on the point was vague. They talked about managing the annual appropriation, but never made the personnel commitment to do so. They talked about Tesseract, but never said what the school budget would look like in five years based on implementing the Tesseract model.

There was an inordinate delay between contract signing and appointing the project manager. Only 15 people, without a formal leader, were on hand to manage a system of mostly underachieving pupils and tenured employees openly hostile to EAI. The board would not cede management control to such an understaffed organization.

Why sould EAI and Hartford both claim that the corporation was the educational manager of the school system? The short answer is that the claim was in everyone's interest. Until the very end, Hartford also benefitted from the unique arrangement. Schools hire consultants every day, but corporate management of schools is a new, innovative national experiment. Favorable local and national publicity can redound to the benefit the city and its leaders. The most direct benefit to Hartford: $1,000,000 plus in capital equipment for city schools. Only corporate ownership, and consequent possibilities for depreciation, made these capital improvements possible.

From the corporation's viewpoint, a simple consultant's contract was not very appealing. Acting as consultant, EAI could not have booked as corporate revenue the entire annual school appropriation, only its consulting fees. As consultant, EAI could not have depreciated equipment. That would have been owned by the school district in the conventional manner. EAI would have looked to all the world like just another small consulting firm offering its services to school districts. It certainly would have appeared to be a step down from the contract already underway in Baltimore. Wall Street might hardly have taken notice.

EAI's innovations were as much in the field of corporate finance as they were in education. But when in 1995 the company decided to direct its management activities to only five schools while still counting the entire annual appropriation in its corporate revenue stream, the jig was up. Even if the board would go along, Hartford city hall would not. In a fiscally dependent school structure, the city holds ultimate financial responsilities. The legal implications of this final "innovation" were simply too frightening.

EAI did not comprehensively manage any of the schools' operations on a day-to-day basis. The role that EAI *did in fact* perform was that of consultant to the board. In that role the corporation was moderately successful. EAI found cost savings, helped the board implement a salary freeze, and pointed a direction toward computer-based education. Ignoring the company's advice, the board chose not to implement a Tesseract system. But consultant advice is frequently ignored; the responsibility for the board's decision cannot reasonably be assigned to EAI.

We can never know how well or badly that system would have worked out in Hartford. But that is not the fault of EAI; corporate management of public schools was never tried. In the absence of any real experience with corporate management of the public schools, the Hartford experiment remains a powerful ideological symbol. For some the experiment will be seen as a victory over downsizing and corporate greed; for others it will indicate a feckless political system and public sector unions grown out-of-control; for the Hartford board the question is what path to pursue now.

IMPLICATIONS

This section of the paper contains personal observations and reflections. The Hartford experiment is discussed within the wider context of the school choice

Table 3.1. School Finance Options

		Producer Competition	
		YES	*NO*
Consumer	Choice YES	#1 vouchers/ tax credits	#2 magnet schools/ charter schools
	Choice NO	#3 EAI in Hartford	#4 neighborhood-based public schools

and other school reform efforts. Prospects for future initiatives from EAI or other private firms are assessed.

The Hartford-EAI Experiment: Privatization Without School Choice

The Hartford-EAI experiment may be viewed as part of the national experimentation fundamentally critical of the way schools are currently organized. Yet the contractual arrangement is unique to date.

Table 3.1 dichotomizes "school choice" into two broad sets of options: choice for the *consumers* of schooling and choice among *producers* of schooling. Choice for either the consumer or producer group alone does not produce choice along full market lines.

Option #4 describes the present system of neighborhood school assignment still in effect throughout most of the United States today. It affords residents of a particular location no choice of schools. Or, saying the same thing, a family's school is mediated by its choice of residence.

Options #1 and #2 have gained the most popular interest and scholarly attention. They represent what is usually meant by the term "school choice," and represent the core of the current public debate. Will there be public school choice only?; or will choice be extended to private schools? This distinction is represented in Table 3.1 as options #1 and #2.

In option #2 there are some school choices open to families; but families must choose among government operated schools. Though terminology may differ, magnet schools are usually the creations of school districts. Charter schools are started by teachers themselves and may take direction from state education departments. But the essential point is this: both magnets and charters are controlled by the state.

Option #1, vouchers, extends government funding to privately organized schools, which often have a history of independence from the state. Parental consumers select from among several competing schools with different organizational sponsors, public and private. Consumers use the voucher to pay for all (or, in some models, a portion of) school costs.

The Hartford-EAI option #3 is unique to date. The board chose EAI from among three bidders (and, effectively, a fourth bidder—the current public school administration). However, all Hartford children continued to attend their neighborhood public school. There was competition among producers of schooling without consumer choice for Hartford residents.

Putatively, this unique arrangement can be explained by special local conditions. Children enrolled in Hartford's Project Concern suburban busing program already did have a sort of magnet schools option. That project was a sore point with some members of the Hartford board because it attracts some of the city's best students. For corporate reasons John Golle wanted to operate all Hartford schools, not just a few. And the Tesseract learning model emphasizes individual choices, not school choices. Perhaps for these reasons school choice for families *within* Hartford was not made an option.

The apparent paradox did not surface in public discussions. School governance structures remained in tact. Hartford's children were still attending public schools, but with a private operator. Whatever the reason, the Hartford-EAI experiment makes researchers and policy analysts aware that there can be a form of privatization without school choice. New forms of school organization can sometimes emerge almost unintentionally, in a political crucible with participants seeking entirely different ends.

Not Cost Effectiveness; Board Satisfaction at a Fixed Cost

Each of the four options in the table provides a different set of incentive structures. Remarks here will be limited to the present case, option #3. Cost-effectiveness models in industry require that both costs and effects vary among competing means of production. Educational applications of cost effectiveness rarely achieve the industrial ideal (Hanusheck et al., 1994; Chubb & Moe, 1990; Wilms & Echols, 1992), but the Hartford experiment doesn't even come close.

Even if the terms of the contract had been fulfilled to the letter, there would have been little empirical basis to judge the value of the experiment. In the Hartford case there were no competing means of production since all schools were EAI schools. Pupils' test scores are the most widely discussed conventional measure of school effects. In Hartford they played no role. By agreement test-score effects were not to be assessed for three years. Even then there were no comparative measures set out in the contract. Virtually the entire contract centered on cost. These could vary below the annual appropriation amount, activating the "savings" clause in the contract. But budgets would grow no faster than a consumer price index. Whether more money could have brought about a positive result would remain an unknown.

As an educational experiment the questions resolved themselves to this: Can cost savings be achieved within the budgeted amount? Will the Hartford board—not the parents—like the result? At best, the EAI experiment would have told us whether the local school board was satisfied with results attained within a given, fixed cost.

Privatization and Choice: Shifting the Debate to the Cost Side

The *Nation At Risk* report in the early 1980s spawned the "school excellence" movement. Since then debates about choice and privatization have been closely linked with school excellence, defined as pupil test scores and other learning outcomes.

The Hartford-EAI experiment had quite a different emphasis, one I have labeled earlier as the conditions of schooling: cleaner, brighter, better-functioning school buildings with more computer-based learning—*and at a cost savings*—formed the basis of the Hartford contract. One suspects that by combing these with a longer school day and year there would be a very wide market indeed.

Pupil learning is educators' turf. Measurement of learning is nebulous and the number of measures available are numerous. Improvements in learning are always debatable. Private schools and private firms possess no special knowledge about pupil achievement. For these reasons emphasis on pupil learning has worked to undermine rather than promote school choice and privatization in the last 10 years. In Hartford John Golle diffused the test score issue swiftly and effectively. To get the contract he simply emphasized cost savings instead. If it works in Hartford why not elsewhere?

Cost control is the private sector's turf. The view is widespread that schooling costs are out of control. In the current environment reaching roughly the same learning levels at a lower cost is more a practicable and more marketable argument for privatization than test score improvement. A clear implication of the Hartford experiment is that future battles over choice and privatization may emphasize schooling conditions over schooling outcomes, cost saving over pupil learning. The latter should be left to professional educators. In this respect the EAI/Hartford experiment provides a positive model for future privatization efforts.

To have actually implemented cost savings, however, three steps would have been necessary in Hartford, and will be necessary if any future, similar efforts are to succeed.

There Must be a Governance Model for Corporate-School Relations

John Golle appreciates the importance of a planned design. In Tesseract he has a tenable educational model. On Wall Street he had an innovative corporate finance model. But he clearly never had a model for school governance and decision making. This has been the chief cause of EAI's setbacks.

Golle lacked even the most basic insights into the differences between politics and markets. Essentially he believed that school boards would act like corporate boards. Show them how to save money and they will do so. This was a tenable belief at the surface; the Hartford board did claim that running an efficient operation is highly important, and they truly did aspire to a management information system which would better help them track money and schooling services. But in the end these were not the board's paramount considerations. In the public sector money saving must be fit in among many other aims. The board's decision to retain hundreds of local employees against the advice of the man from Minneapolis was a predictable one.

Because the board's reaction was so predictable, the convention of designating all school revenues as corporate revenues must be questioned in this case. There is much to admire about John Golle, including his vision of the public interest. Too many locals saw Golle as, "The Music Man come to River City" or "the white folks trying a con job."

For any model to be comprehensive, the corporate community, as well as the educational community, local voters and schoolchildren must be aware of the aims each has for the other. This was not the case in Hartford.

The Private Contractor and its Public Employees Must Work Cooperatively

The Hartford experiment was a classic example of an innovation described in organization behavior textbooks as "top down." The board imposed it on the system. School employees were unified in their opposition. A mountain of studies and research projects presaged the outcome of this experiment: employee opposition alone likely would have killed it over time, even if other factors had been favorable. EAI only had fifteen employees in Hartford. More would not have helped, and would have been too costly in a project ultimately designed to save costs.

For corporate management of the public schools to work, corporations and public sector unions must negotiate together. This may not be such a far-fetched prospect. Even in a highly unionized state like Connecticut, teacher unions are coming under increasing criticism. Nationally, public employee unions are rapidly spending their political capital in efforts to retain their membership's educational monopoly.

Suppose at some future stage unions saw *their* school choice as one between a Hartford-EAI model and a fully privatized, voucher-based model with no unions. Unions would never be joyous at the prospect of their own downsizing. But at that point unions might be willing work with corporations in achieving a smaller, better paid membership—better paid through technological efficiencies. In any case, "over my dead body," won't work. The dead bodies will always be the corporations.

The Private Contractor and the Board Must Have a Clear Understanding of Plans for Downsizing Over the Period Specified

This brings us to the third and most crucial point. In Hartford neither party—the board or EAI—fixed their attention on a long-range plan. In such an environment it is easy to understand how misconceptions could arise. Board members could reasonably have thought they were voting for free computers; the company could have thought the board would ultimately endorse their Tesseract model.

This should not happen again in other localities. During the proposal stage, a corporation should lay out its instructional and expenditure plans. If it does not do so, the board should insist on these data, and should compare plans of multiple vendors. Because cost savings are only one consideration among many others in the public sector, it behooves both the board and the contractor to clarify their plans fairly precisely. Lowest cost is not always the appropriate educational issue. But cost and instructional plans would help voters focus on those issues during school board election campaigns.

A Final Note For School Finance Specialists

At many points in the story corporate developments controlled events and influenced outcomes. Golle's insistence on being school manager rather than a school consultant reflected his desire to show sales growth to his investors. The computer-based, Tesseract learning model had distinct possibilities for helping the corporate income statement. And the corporate deficits announced in December help explain his withdrawal from the city even though his slate had just achieved a school board election victory. These are only three very clearest examples. There are doubtless many other points at which corporate considerations impacted educational events.

The field of school finance traditionally has relied on public finance, law and political science as its three main allied disciplines outside the field of professional Education. Hereafter, these allied specializations may not be enough. As privatization and choice issues come to the fore, we must become knowledgeable in fourth allied field: corporate finance. The intersection of corporate finance and public policy in education seems a fruitful area for further research.

NOTES

1. Biographical material on John Golle and the Tesseract Development Company was gathered via Lexis/Nexis from the following sources: *Minneapolis Star Tribune,* June 4, 1994; *The Washington Post*, March 13, 1994.

2. Material on the Tesseract Development Company and EAI's private school operations were gathered via Lexis/Nexis from *Public Relations Newswire*, the Tesseract Development Company, January 5, 1995; *Software Magazine,* January 1995. "Editorial—Saturday Mail Box" *The Baltimore Sun,* Saturday, December 24, 1994. *The Washington Post,* March 13, 1994. On December 31, 1994 EAI bought outright the private schools in Scottsdale and Paradise Valley from the Tesseract Development Corporation.

3. John M. McLaughlin, *Education Industry Report,* vol. 3, no. 8., September 1995, p. 7.

4. Dow Jones, Electronic News Service, *//quick EAIN*, company disclosure forms.

5. See the *Value Line Investment Survey.* "The Value Line" is defined as follows: "Reported earnings *plus depreciation* ("cash flow") multiplied by a number selected to correlate with the stock's target price projected out to 1998-2000." Value Line does not follow EAI; however, Value Lines measures are widely used by other individuals and firms to evaluate common stocks. See also Merrill, Lynch, Pierce, Fenner and Smith, 1974.

6. For a discussion of this problem at the national level see Bruce Cooper and associates, 1994.

7. The Hartford school board is typical of the nation in this respect. See, for example, Wirt & Kirst, 1992; Vitaritti, 1983; Lutz & Iannaccone, 1978; McCarty & Ramsey, 1971; Dansberger, Kirst, & Usdan, 1992.

8. Just before the board decision the teachers union distributed a study deeply critical of test-score performance in EAI's Baltimore schools. Arguably, attitudes hardened in the face of this kind of pressure, but the board could not appear to cave in.

9. Much of the narrative here is based on the contract document itself. See "Contract For Professional Services By and Between Board of Education of the CITY of Hartford and Education Alternatives Inc.," First Complete Draft, September 22, 1994.

10. The provision that the schools' annual budget would grow no faster than the Consumer Price Index was eliminated from the final contract.

REFERENCES

American Federation of Teachers. (1994). Corporate report Minnesota, June, 1992. In *The Private Management of Public Schools, An Analysis of the Chubb, J.E., & Moe, T.M. [1990]. Politics, markets and America's schools.,* Washington, DC: Brookings.

Cooper, B. et al. (1994, Summer). Making money matter in education: A micro-financial model for determining school level allocations, efficiency and productivity. *Journal of Education Finance, 20* (1), 66-87.

Dansberger, J.P., Kirst, M.W., & Usdan, M.D. (1992). *Governing public schools: New times new requirements.* Washington, DC: Institute for Educational Leadership.

EAI Experience in Baltimore. Washington, DC: American Federation of Teachers.

EAI the first year. (1995, June 25). *The Hartford Courant,* p. A4.

EAI The Hartford experiment. (1996, January 24). *The Hartford Courant*, p. A10.

Hanusheck, E.A. et al. (1994).*Making schools work.* Washington, DC: Brookings.

Lutz, F.W., & Iannaccone, L. (1978). *Public participation in local school districts: The dissatisfaction theory.* Lexington, MA: D.C. Heath.

McCarty, D.J., & Ramsey, C.E. (1971). *School managers: Power and conflict in American public education.* Westport, CT: Greenwood.

Merrill, Lynch, Pierce, Fenner and Smith. (1974). *How to Read a Financial Report.* New York: Merrill, Lynch, Pierce, Fenner and Smith.

Stecklow, S. (1995, May 2). Florida pact to end for education alternatives inc." *Wall Street Journal.*

Vitaritti, J.P. (1983). *Across the river.* New York: Holmes and Meier.

Wilms, D., & Echols, F. (1992, December). Alert and inert clients: The Scottish experience of parental choice of schools. *Economics of Education Review, 11* (4), 339-350.

Wirt, F., & Kirst, M. (1992). *Schools in conflict: The politics of education.* Berkeley, CA: McCutchan.

Chapter 4

FREEING RESOURCES FOR IMPROVING SCHOOLS: A CASE STUDY OF TEACHER ALLOCATION IN THE BOSTON PUBLIC SCHOOLS

Karen Hawley Miles

Increasingly, discussions of education reform end up in debate over how much money will be required to fund it. While educators often argue for more funds and critics counter that school spending is already at an all-time high, neither group has looked systematically at the use of existing resources. This analysis addresses how school systems might reconsider their use of teachers, their most important and expensive resource, to support reform. A case study of Boston Public Schools shows how the specialization and fragmentation of teaching

Advances in Educational Productivity, Volume 6, pages 67-97.

ISBN: 0-7623-0082-5

resources reduce the individual attention most students receive and limit a school's flexibility to respond to student needs. The analysis develops three measures that might be used as indicators of the size of opportunity for freeing teaching resources. It identifies four practices that together account for over 40 percent of Boston's teaching resources and could provide significant opportunity to redirect resources to support reform.

BACKGROUND

Education researchers and reformers increasingly agree that improving schools demands changing the way teachers work in schools. Like any organization where quality depends on the knowledge, capability, and commitment of its professionals, schools must find time for teachers to learn new roles, keep pace with advances in their profession, and share insights with colleagues. Further, many educators support changes in practice which demand teacher collaboration in creating curriculum, diagnosing, assigning, and instructing students. At the same time, many reformers argue that more flexible ways of organizing instruction could respond more effectively to differences in learning needs.

This need for time and flexibility has become a seemingly insurmountable barrier to change, as educators often argue that it will cost more money and critics accurately counter that spending on teaching has never been so high (Peltzman, 1993; Finn, 1993). Why is it so difficult to free teaching resources when schools average one classroom teacher for every 17 students nationwide? (NCES, 1992) This case study explores this question by detailing how Boston Public Schools allocate classroom teachers, outlining alternatives to the current organization, and describing the underlying educational and management practices that make new ways of organizing schools so difficult.

METHODOLOGY AND ORGANIZATION

The Boston Public Schools (BPS) provide a useful site to study the allocation of teachers for three reasons. First, Boston spends as much for each pupil as many urban districts undergoing reform. Second, most schools in Boston allocate teachers in traditional ways, assigning one to a classroom, using separate programs for students with special needs, and following the familiar five to six period daily schedule at the secondary level. However, at the same time, a number of Boston schools—such as Lyons, O'Hearn, Jackson Mann, and Timilty—have reorganized to provide more flexible, individual instruction to all students and to provide more planning time for teachers. The experiences of these pioneering schools highlight alternatives to the status quo and help to identify the challenges to changing the way teachers work in schools.

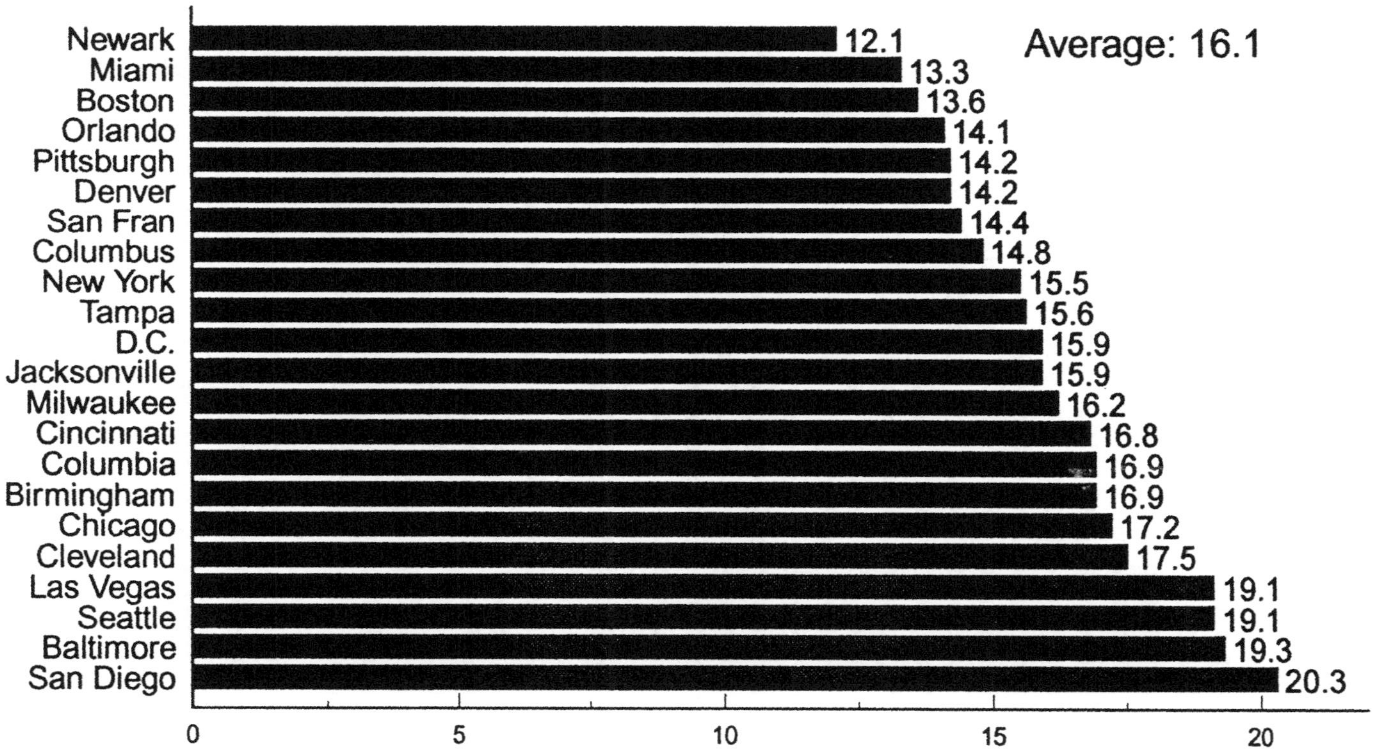

Source: Council of Urban Boards of Ed. 1989 data; urban districts with over 40,000 enrollment.

Figure 4.1. Boston versus Urban Schools Pupils per Teacher Fiscal Year 1989

In 1990-1991 Boston spent $440 million in city, state, and federal funds for each of 55,000 students, or nearly $8,000 per pupil. The National School Board Association (NSBA) survey of 1989 spending, one of the only reports that include all sources of funding, shows total urban spending ranging from $3,774 per pupil in Birmingham, Alabama to $9,700 in Chicago. Other districts spending near $8,000 per pupil include Milwaukee, at $8,922, Pittsburgh at $7,266, and Newark at $7,394 (Council of Urban Boards of Education, 1990, pp. 113, 9). Like most districts, Boston allocates close to 60 percent of all spending to classroom teachers.[1] In 1991 Boston had one classroom teacher for every 13 students in grades 1-12, not including teachers working in administrative positions outside of schools. Analysis of NSBA data shows that in 1989 urban districts with more than 40,000 students averaged close to one teacher for every 16 students. Figure 4.1 shows that staffing ratios ranged from 12.1 students per teacher in Newark, to 20.3 in San Diego. Using data from the National Center for Education Statistics (NCES), Picus finds that school districts like Boston, whose student populations include more than 50 percent minorities, averaged 16.6 students per teacher in 1987-1988 (Picus, 1993).

At first blush, these numbers seem to contradict the much publicized descriptions of overcrowded classrooms and urgent cries for more teaching resources. But, in interviews, Boston principals and teachers cite more teaching resources as their most pressing need. To understand this apparent conflict, this analysis traces the difference between the district-wide average in Boston of 13.2 students per teacher and a much different reality which students and teachers face. Actually, most students spend the majority of their time in classes well over 23, almost twice as large as the theoretical potential of 13 would indicate. Regular education class sizes average 22.7 and vary substantially around this average from 10 to 33. Teachers have fragmented and uncoordinated time available for planning. And, most teachers at the secondary level have responsibility for over 100 students per day, a stark contrast to the apparent potential for flexible, individual instruction that the ratio of 13.2 students per teacher would suggest.

Four managerial and educational practices explain the difference between the theoretical potential of 13.2 students for each teacher and these realities:

1. Large numbers of specialized teachers working outside regular classrooms with specifically defined classifications of students.
2. The practice of providing teachers with planning time in short, fragmented periods during the school day and using other classroom teachers to cover instruction at these times.
3. A formula-driven approach to grouping students for instruction that is guided by formal student classification and contract guidelines rather than school-level decisions.

4. The fragmented daily schedule at the secondary school level in which teachers instruct five completely different groups of students for less than an hour each day.

Changing any one of these practices would free incremental teaching resources, but rethinking all of these practices at once could create opportunities for true reinvention of schools.

The following pages describe how each of the four organizational practices outlined affect the allocation of teachers and limit opportunities for more flexible use of resources. Four sections will:

a. Describe current practices and quantify their effect on class sizes and teacher loads;
b. Identify alternatives and quantify how student teacher ratios and teacher loads could change given new assumptions; and
c. Highlight barriers to changing these practices.

Four strands of analysis contribute to this synthesis of opportunities. Accurately portraying the number of teachers working in classrooms required working with BPS to adjust traditionally reported numbers in two ways. First, rather than reporting only the teachers paid out of general funds, we added all teachers regardless of funding source. Second, we isolated those teachers working with students in schools from those working in administrative positions, excluding the latter from this analysis.[2] Next, school-level staffing reports for each of BPS's 115 schools allowed examination of teacher allocation by school, grade, and program. All enrollment, staffing, and financial data represent the 1990-1991 school year unless otherwise noted. Third, extensive interviews with district administrators and 20 school principals highlighted alternatives to traditional school organization and barriers to change. Finally, analysis of the 1989-1992 BPS union contract and the union grievances related to changes in school staffing pinpointed contractual barriers.[3]

Throughout this paper, I will calculate the percent of teaching resources "freed" by changing a given practice and the resulting effect on average class sizes. In doing so, I do not mean to suggest that all classes should be one size, or even that smaller classes are necessarily better in every case. Instead, I use these ratios as a short-hand way to measure the relative size of the opportunity to free resources for new ways of organizing schools. As Seymour Sarason wrote:

> The fact is that one of the major factors maximizing the gulf between educational goals and accomplishments has been the way resources have been defined.... There is a universe of alternatives one can consider and if we do not confront that universe, it is largely because we are committed to a way of defining who should be in the classroom.... One teacher to one classroom is not an end in itself, but one means to providing more time for individual students when needed (1982, pp. 275, 284).

Schools might choose to use these resources in many different ways, such as lowering class sizes, implementing team teaching, creating more teacher planning time, or purchasing instructional equipment.

STUDENT AND TEACHER CLASSIFICATION

Current Allocation of Teachers

Boston's "pull-out" approach to serving special needs students largely explains why regular education class sizes average 23 despite a district-wide average of 13 students per teacher. Thirty percent of Boston's students are classified as having a learning need requiring specialized attention. As Figure 4.2 shows, over 40 percent of BPS' teaching force works outside the regular classroom to serve these students through the special education, Chapter 1, and bilingual programs.

Special education teachers, constituting 23.5 percent of the teaching force, represent the largest portion of teachers working outside the regular classroom. These teachers work with the 22 percent of students in Boston who are categorized as having some learning or physical condition requiring "special education." By law, each of these students must receive appropriate support in the "least restrictive environment" possible. Though the law was intended to encourage "mainstreaming" of special needs students, BPS primarily uses a "pull-out" model. In this model, students with milder disabilities, nearly 14 percent of all BPS students, leave the regular classroom for portions of the day to work with specially certified resource teachers individually or in groups no larger than 10. Students with more severe behavioral, educational, or physical disabilities, 8.5 percent of BPS students, work with appropriately certified teachers in separate classrooms of no more than 12 students.[4]

Despite these efforts to classify and separate students according to their needs, even students in small, self-contained classrooms do not necessarily have the same requirements. For example, a class of 10 "Learning Adaptive Behavior" students, the most common classification in Boston, might all have difficulty paying attention in class, but the source of their difficulty may be different. Some students may find the pace of the regular classroom too slow while others may have problems keeping up. Thus, the smaller group may make it easier for the teacher to develop a customized education plan, but it does not eliminate all of the diversity in student learning needs.

Bilingual teachers, the next largest group working outside of the regular classroom, comprise 13 percent of the teaching force. Over 7,000 students in Boston receive bilingual education in separate, self-contained classrooms in one of ten different languages. These classes also have lower mandated class sizes than regular students, capped at 18 per teacher or 25 with a teacher and aide.

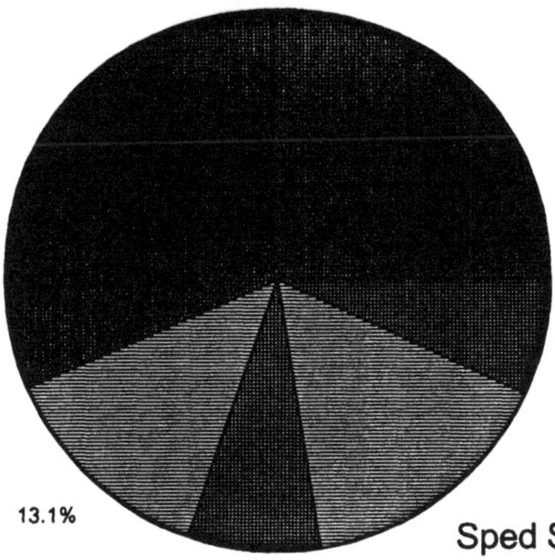

Source: 1990-1991 Staffing Reports.

Figure 4.2. Percent of Teachers by Type

Chapter 1 teachers, constituting 6.9 percent of the teaching force, represent the final group working outside the "regular" classroom in Boston. Schools with students from families earning below the poverty level are eligible for federal Chapter 1 funding. Where 75 percent of the students in a school qualify, the school may use these funds to support the whole school. The majority of Boston schools qualify to use Chapter 1 funds in this way. However, Boston has traditionally used the "pull-out" model, in which eligible students leave regular classrooms to work with resource teachers, largely in reading and writing.

Implications and Alternatives

Using so many teachers outside regular classrooms means that all students, except those receiving completely separate education, spend the bulk of their time in larger instructional groups. In addition, Boston, like many schools systems, has needed to develop administrative systems to monitor and coordinate the various teachers and service providers working with special students. This process diverts even more resources from the regular classroom, gradually making them larger. So, ironically, the very action designed to lessen

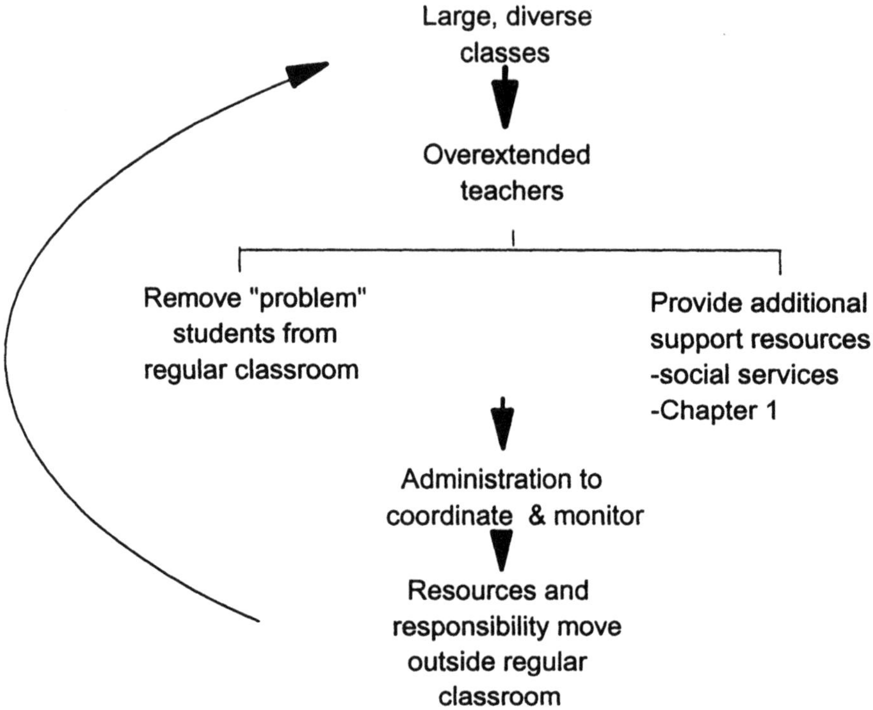

Figure 4.3. Cycle of Specialization

the demands on teachers, referring a student outside the regular classroom, may increase them. As Figure 4.3 shows, the pattern is self-perpetuating, the more schools designate specialists to respond to diverse learning needs, the less resources and pressure exist to help teachers in the regular classroom develop these skills.

This cycle may help explain why the proportion of students classified as requiring special education in "substantially separate" classrooms had risen three times faster (at 5% a year) than any other class of students from 1987 to 1991. This growth in students who spend almost no time in "regular" classrooms occurred at the same time that the portion of mainstreamed special education students, those students who spend most of their time in regular classrooms, declined at nearly the same rate. Table 4.1 shows the growth rates by class of students.

Given that 43 percent of Boston's teachers work outside the "regular" classroom, numerous alternatives exist for creating smaller, more flexible groups for all students. Several BPS, such as Lyons and O'Hearn have combined all of their special needs students and teachers into heterogeneous

Table 4.1. Annual Growth Rate of Students by Clasification

Student Classification	*Average Annual Percent Growth (1988-1991)*
Regular Education	+1.5%
Mainstreamed Special Education (.1 to .3)	-5.1%
Separate Special Education (.4)	+5.8%
Bilingual	+2.4%
Vocational Education	-1.9%
Total	+1.1%

Source: Boston Public School Enrollment reports 1988-1991.

classes smaller than 15. Others, like Jackson-Mann elementary and Timilty and Taft middle schools have integrated only the resource teachers to create smaller student groups. At O'Hearn, a team of one regular education teacher and one specialist work together with a heterogeneous group of 20 students. As a team, the teachers group students in different ways throughout the day depending on the lesson. At James Curley Elementary School, all students and teachers divide into sections of 15 or less for 90 minutes of reading each day. As these examples show, though some schools use integration to create smaller class sizes, others use the flexibility to adjust resources to match the student and lesson.

The Boston case allows quantification of the resources pull-out practices draw away from regular classes and provides a sense of the relative impact of segregating different groups of students and teachers. If all Chapter 1 and special education students returned to the regular classroom, the district-wide ratio would fall by six, from 22.7 students for every teacher to 16.6. Tracing this potential by program, the district-wide average of 22.7 would fall to 20 with the addition of Chapter 1 teachers into regular classrooms, to 17.8 with the integration of resource teachers, and to 16 if all substantially separate teachers and students moved into heterogeneous groups.[5] These groups of 16 students would include one special education student formerly in a "substantially separate" and two who previously left the classroom for a few periods each week.

As Figure 4.4 shows, the impact of pull-out differs by school level depending on the mix of programs. Integrating any single group of classified teacher or student into regular classrooms might not free enough resources to alter teaching conditions. But, by viewing all of these teachers together, rather than treating them as part of isolated programs, new possibilities emerge.

A significant and growing body of research supports the use of smaller, more inclusive groups of students organized in more flexible ways. While research on class size and educational achievement does not show conclusive gains in student performance from small reductions, it does suggest improvement when class sizes drop to a threshold small enough to enable changes in teaching

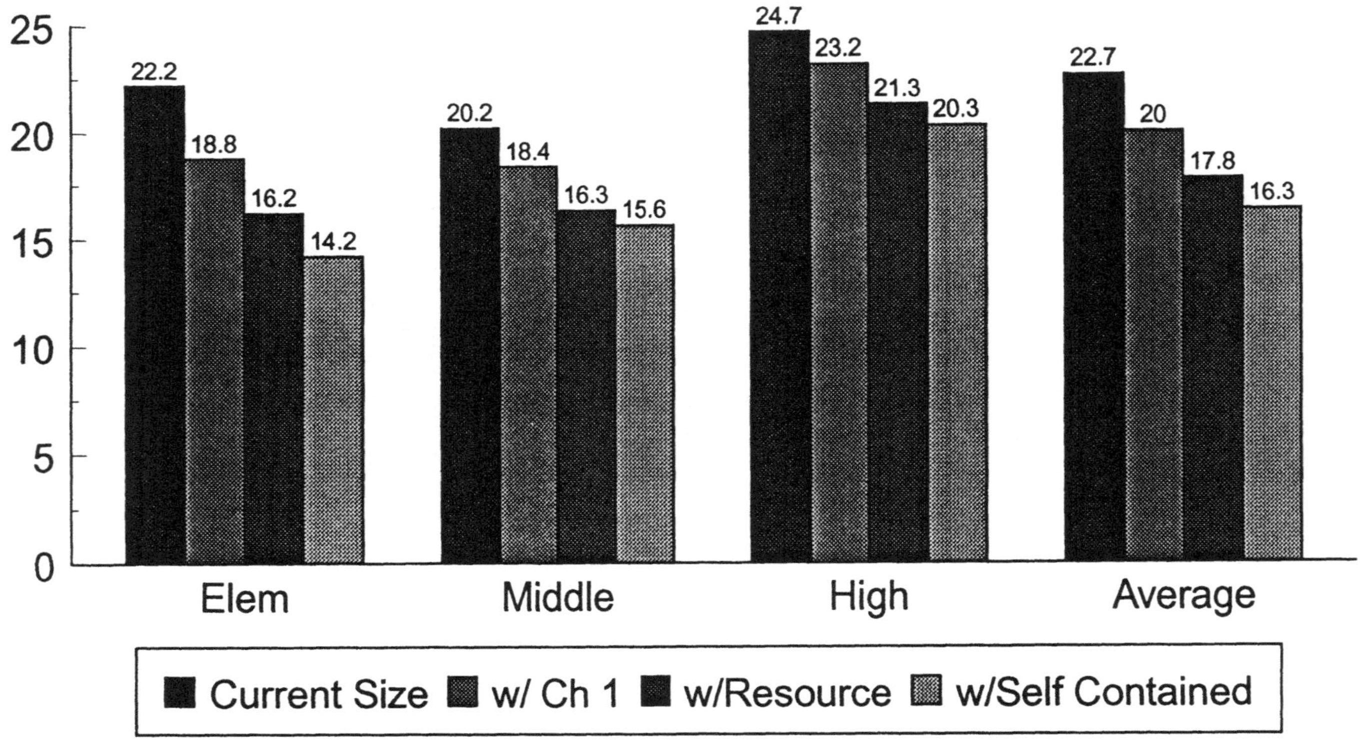

Figure 4.4. Impact of Integrating Specialists Pupils Per Teacher

practices and more individual attention (Hanushek, 1991). Estimates of this threshold level vary from the low 20s (Ferguson, 1991) to the teens (Folger, 1992; Glass, 1979; Levin, 1988; Nye, 1992). Further, research on the effectiveness of pull-out programs for students with special needs shows that students labeled as "special" may have less self-esteem, less confidence in themselves as learners, and suffer from lower expectations of teachers as well. From a practical standpoint, pull-out practices disrupt a student's participation in the regular classroom, making it even more difficult for these students to contribute (Cuban, 1989; Levin, 1987; Oakes, 1985). At the same time, research demonstrates that creating more inclusive groups does not necessarily lower the achievement of more accomplished students (Skrtic, 1991; Slavin, 1991; Wheelock, 1992).

Researchers do not suggest eliminating all instructional grouping. Instead, they argue for a more fluid approach. Groups would be formed according to interest and learning styles as well as differences in performance. Any student assignment based on achievement or special learning need would be short in duration, focused on the development of specific skills and aimed at reducing or eliminating achievement gaps (Common Destiny Alliance, 1992). The acclaimed "Success for All" model has significantly raised student performance using this approach. The model shifts resources from Chapter 1 and special education to support intensive instruction in reading for students in the earliest grades (Slavin et al., 1994).

Barriers to Reducing Student and Teacher Segregation

Chapter 1 and special education legislation establish the framework for the formal classification of students according to learning needs. They also require that schools give additional support to students with certain learning or demographic characteristics. But this legislation does not require the use of pull-out programs. Yet, few Boston schools have broken through the complex web of tradition and practice that support student segregation and teacher specialization. Those schools that attempted it speak of four sets of obstacles.

1. Current funding mechanisms and district staffing practices encourage the classification and separation of students.
2. Parents of both regular and special education students sometimes lobby for separate placement fearing that changes will reduce the attention their children receive.
3. Effectively meeting the needs of regular and special education students in the same classroom can require teachers to change their teaching practice and develop new pedagogical and diagnostic skills.
4. More flexible, integrated schools challenge the traditional structure of schools and demand new organizational abilities.

At the highest level, Chapter 1 and special education legislation allow more flexible use of resources. The majority of BPS elementary and middle schools meet federal criterion allowing them to use Chapter 1 funds school wide. Therefore, if schools can show how using Chapter 1 funds with smaller, heterogeneous groups of students benefits targeted students, they may do so within regulations. However, many districts continue pulling students out for Chapter 1 services because they can directly demonstrate compliance with the stipulation that the funds supplement not supplant existing resources. Similarly, even though the more severe categories of special education students typically receive education in separate settings, schools may mainstream most of these students if parents and educators agree. According to Public Law 94-142, each student classified as requiring special education must have an "Individual Education Plan" (IEP) developed and approved by the student's parents, teachers, principal, and specialists. The BPS that have integrated their special needs students, such as Lyons and O'Hearn, have shown parents how the new structure will more effectively meet their child's needs.

While the law allows schools to use special education and Chapter 1 resources in the regular classroom, schools have a strong incentive to continue the classification of students for the purposes of garnering more funds. Schools receive extra staff based on the number and mix of special needs students they have. The more students that a school has classified as requiring separate instruction, as opposed to mainstream instruction, the more staff they receive. The district receives partial reimbursement from the state based on the number of special needs students, but these funds do not begin to cover the full cost of special education. Thus, the school-level incentive diverts resources from regular education. Boston is working to find a balance that provides extra resources to schools with needier students but at the same time encourages schools to serve these students in the most cost-effective and beneficial way. BPS has begun to monitor the numbers of students referred to special education and to hold schools accountable for limiting segregation to students who might truly benefit from separate instruction. In addition, the 1994-1997 BTU contract contains class size guidelines which give schools more teachers for serving special needs in the mainstream.

Gaining the parental support required to implement inclusion can be a time-consuming and frustrating effort. Not only must parents of students with special needs agree to the strategy, but parents of students in the regular classroom must understand and support it. At one Boston school, full integration of resource teachers and students into regular classrooms generated such dissension that a group of parents switched their children to other schools. Both O'Hearn and Lyons, which include a number of typically segregated students in each classroom, embrace the philosophy that integration improves learning opportunities for all students. Through BPS' choice plan, these schools actively recruit parents supportive of their approach. They hold frequent workshops and open houses to educate and include parents.

While smaller groups of students create the conditions in which a teacher can respond to more diverse student needs, teachers of integrated groups need to draw on a wide pool of knowledge, curriculum, and pedagogy to engage the full range of students capabilities. As one principal put it:

> the biggest mistake is to assume that smaller classes automatically solve all problems dealing with diversity. Once you have integrated classes, you've got to face the fact that you've fundamentally changed your approach to schooling... and even if you can get this right... all it takes is one student acting out to destroy the most engaging, sophisticated lesson.

The Boston schools working to include more special needs students speak passionately of the need for time and support to develop new curriculum, practice new teaching strategies, and work with specialists.

Finally, whether the school implements a team teaching model like O'Hearn, or uses specialists in a more consultative role, reducing separation and specialization demands more collaboration than a pull-out model. Not only does this require teachers to develop new organizational skills (Gartner & Lipsky, 1989), but it challenges the traditional structure and culture of schools in which specialized teachers work individually with similar groups of students (Skrtic, 1991). In a historical analysis, Tyack and Tobin argue that reforms which seek to change this culture are "messy, difficult... and require intense and continual public dialog about the ends and means of schooling" (Tyack & Tobin, 1994).

At the heart of reconsidering the use of pull-out programs lie two fundamental educational and organizational decisions about how to accommodate diverse student needs. First, should differences in learning needs be treated as exceptions? And, is specialization the best way to organize to serve such differences? This analysis highlights the cost of specialization of teachers and categorization of students. If a school district has very few students classified as needing specialist attention outside the classroom, these students can be served without diverting significant resources from other programs. But, the Boston case shows that as the numbers of "special" students grow, this diversion becomes significant. When over 40 percent of teachers work with "special" students, it may be time to reconsider the definition of "special," especially given the mixed evidence that these practices improve achievement.

PLANNING AND PROFESSIONAL DEVELOPMENT TIME

Current Allocation of Teachers

The practice of hiring additional teachers to cover teacher planning time accounts for the remainder of the difference between the unadjusted district average of 13 and current class size average of 23. In most school districts, teachers have some time during the school day when they do not instruct. In Boston, elementary school teachers have four, 45-minute periods a week (three in 1991) to use as "planning and development" (p&d) time. At the secondary level teachers have two periods of seven during which they do not teach, or 30 percent of the instructional day. A teacher may be assigned administrative duties during one of these periods. In addition, teachers have time for lunch. Typically, these breaks from instruction are scheduled randomly throughout the day. Like most school districts, BPS uses certified teachers to cover classrooms during this time. At the elementary school level this totals 142 "specialists," or 9 percent of elementary school teachers, who teach art, music, gym, and computer classes. At the secondary level providing teachers with administrative duty and p&d time requires Boston to hire 30 percent more teachers.

Implications and Alternatives

Especially at the secondary level, noninstructional time consumes significant teaching resources. But, the short, fragmented blocks of noninstructional time in the current structure provide opportunity for little more than brief breaks, clerical work, or quick review of an upcoming lesson. The time does not exist at the elementary level and is not designed at the secondary level to allow more substantive planning and collaboration. These activities require longer blocks of uninterrupted time which is coordinated with other teachers.

Through careful scheduling and coordination, a number of Boston schools have created longer blocks of individual planning time by combining two periods, releasing teachers from administrative duty, and sometimes adding lunch or after-school time. But the current practice of using certified, mostly full-time teachers to cover breaks in instruction, makes creating sustained blocks of common time difficult. Teachers who cover instruction during this time do not get to participate in "common" planning time. In addition, scheduling planning time for groups of teachers at the same time generates the need for coverage of instruction at concentrated points in the day. For example, a specialist in a small school who covered five teacher planning periods over the day could not handle all five periods at once.

Boston schools have used or proposed two sets of alternatives to overcome these difficulties; using new staffing arrangements to cover common planning time or not providing teacher-directed instruction during this time. One way to organize staff to cover peaks in demand would be to hire part-time teachers. Another solution proposed by several Boston elementary schools, would be to use the funds for specialists to contract with providers outside the school for music, art, language, or computer instruction. Smaller student to teacher ratios generate additional opportunities to regroup teachers and students during the day to find new time for teacher planning and development. For example, at the Kent School, teachers occasionally combine two small classes, creating instructional groups of 30 during which one teacher might hold parent conferences or work on curriculum. Several secondary school principals suggested that creating substantially larger class sizes for nonacademic subjects such as physical education or music would free resources for planning. For example, with student groups of 16 instead of 24, schools might consider combining two groups to create physical education classes of 32.[6]

This more flexible grouping of students would create individual or group p&d time, but it does not allow time for teachers to work together in teams or as a school. Doing this without using additional teachers to fill the instructional time requires either filling student time in other ways or having teachers plan after-school hours. Groups of teachers and administrators in most Boston schools voluntarily meet regularly after school to plan together. But, they add this time onto their workdays rather than rethinking the structure of time. In 1992 Boston's North Zone administrators and principals proposed an alternative schedule that maintained the same instructional time for students but created two hours every other week for school-wide planning. In this schedule, students would be released from school an hour early every other week. To make up for this time, administrators proposed extending each school day by a few minutes each day. They modeled their plan after Newton, Massachusetts, which releases students one hour early twice a week.

Like pull-out practices, the current structure of p&d time is in part an incremental response to large classes and inflexible student grouping and scheduling practices. Large classes increase the demands on teachers, creating the need for breaks. While class size restrictions and tradition limit the options for regrouping students to provide such time. One reason that secondary school teachers have so much more p&d time than their counterparts in elementary schools is that p&d time limits a teacher's student load. In the current system, teachers typically work with five classes a day. At current class-size averages, this translates into a load of 100 students per day at the middle school level and 125 in high school. Smaller groups of students alter the assumptions supporting current practice.

Rethinking the structure of time, teachers might find that a different allocation of resources would increase their ability to provide individual attention and create more usable time for planning. As Figure 4.5 shows, reconsidering the use of p&d time along with pull-out practices frees 42 percent of all teacher time for alternative configuration.[7] The 58 percent of time remaining represents the portion of teaching resources that would be needed to provide all students with the classroom resources regular students currently receive.

Using all of this time to reduce class sizes would create groups of 13 for all students. So, for example, secondary school teachers might choose to forego one planning period to reduce the size of each class and use options discussed

	Elementary	Secondary	Total
Regular Class Time (All students at current regular ratio)	59.4%	56.8%	57.9%
P&D time	8.8%	30.0%	20.7%
Chapter 1 Class Time	8.2%	3.7%	5.6%
Special Education Resource Room Class Time	8.0%	5.1%	6.4%
Incremental Class Time for Separate Special Education	11.3%	4.1%	7.3%
Incremental Class Time for Separate Bilingual Education	4.3%	.3%	2.1%

Figure 4.5. Teaching Resources Freed
Percent Resources for P&D and Extra Support to Special Students

earlier to add to their planning time. Because elementary school teachers receive only four, 45-minute breaks from instruction a week, the change here appears more minimal. But, looked at another way, reallocating this time could still be significant. In most Boston elementary schools, providing this time requires hiring one or two additional teachers. Redeploying just one of these teachers to the first-grade regular classrooms could reduce first-grade student teacher ratios in most Boston elementary schools to fewer than 14.[8]

Barriers to Reconsidering Planning and Development Time

These alternatives for making existing time more effective for planning and development include at least five possible changes:

1. Combining planning periods to create longer blocks of time
2. Specifying that teachers use some of this time for common planning time
3. Using noncertified, contracted, or part-time teachers
4. Regrouping students, in some cases exceeding class size maximums, in certain subjects or for particular lessons
5. Extending the teacher workday

Each of these proposals would require significant changes in the BTU contract as well in the spirit with which both the union and the administration approach the issue of teacher workday. Like many unions across the country, the BTU has viewed issues like defined planning periods, class size maximums, and hours worked as fundamental to making the job of teaching more humane and professional (McDonnell & Pascal, 1988). Accordingly, the BTU has traditionally opposed any efforts that might infringe on these hard-won provisions.

Controversy over elementary school p&d time illustrates the difficulty of changing the structure of p&d without union collaboration. Any alternative staffing arrangements such as the use of part-time teachers or contracting for instruction must be negotiated through the BTU since these strategies could result in lost teaching jobs. School-based management teams that have petitioned the BTU for waivers to use "specialist" resources in other ways have been denied. The BTU contract and a 1991 grievance settlement limit school-level flexibility in scheduling a teacher's workday. The contract entitles elementary school teachers to four periods of 45 minutes per week to use as they wish. Administrators may not structure teachers' p&d time. This means that principals may not require teachers to use it for team planning or other coordinated activities. The 1994-1997 contract adds one period a week that can be used for common planning time. The workday for specialists and generalists assigned in elementary schools must be divided into six periods of 45 minutes each. No other schedule is permissible without a waiver. Class size restrictions are to be strictly enforced, prohibiting such innovations as gathering students in large groups for certain activities, followed by individualized attention in small clusters.[9]

Thus, current precedents in Boston suggest that unions might resist efforts to reconsider the way in which schools provide teachers with p&d time. But, these resources represent 30 percent of all teaching resources at the secondary school level and 9 percent at the elementary school level. Given the existing system, with large classes and unmanageable student loads, the current arrangements may seem logical. But, changing these assumptions, creating much smaller classes and rethinking how the school day is organized, opens potentially more effective options.

STUDENT ASSIGNMENT TO SCHOOLS, GRADES, AND PROGRAMS

Current Allocation of Teachers

To this point, I have focused on explaining why the district average of 13 students per teacher does not equal the average regular education class size of 22.7. This discussion misses the reality, described earlier, that regular class sizes vary widely around this average. Eliminating unplanned differences in class size could represent another opportunity to redirect teaching resources. To explore this possibility, I sought to understand how often class sizes in the same program differ from the average and why.

Like most school districts, Boston allocates teachers to schools by formula. The district uses past enrollment combined with demographic trends to project the number of students for each school by grade. Schools receive the required number of teachers to stay within class size boundaries for each grade. The BTU contract caps regular education elementary school classes at 28, middle school at 30, and high school at 33. The contract specifies lower maxima of 20 for occupational education classes like home economics and industrial arts, reasoning that they require closer supervision of students using specialized equipment. So, in a simple example, an elementary school projected to have 50 regular education students in the first grade, would receive two teachers for that grade, or the closest whole number to 50 divided by the class size maximum of 28.

Thus, given the allocation formula we would expect regular education class sizes to vary mostly based on school level and subject. The first question to ask is whether the formula reflects a conscious educational strategy. And, when applied, does the formula work? That is, does the formula result in staffing that meets the educational goals it intends? Does it result in equal distribution of resources across schools and grades? Does it have unplanned side effects?

The staffing formula incorporates a deliberate strategy to concentrate resources in the earlier grades. It also intends to adjust resources depending on each school's mix of students. BPS does not completely accomplish these goals. The formula aims to create the smallest class sizes at the elementary level and the largest in high schools. But, class sizes average 22.2, 19.6, and 24.7 at the elementary, middle, and high school levels respectively. Thus, the gap between the maximum class size allowed and the actual average is 6 at the elementary school level, 11 at the middle school, and 8 at the high school level. Further, at each level, class sizes for students in the same grade and program may vary widely around these averages.

Even with strict application of the formula, Boston's many small schools combine with the fragmentation of schools into distinct age groups and small programs within schools to make it difficult to control variation in class size. An analysis of elementary school class sizes provides the simplest example of this phenomenon. With class sizes capped at 28, when the 29th student enters a school with only one class in that grade, a new teacher must be added. Thus, the class size average falls dramatically from 28 to 14.5. A similar pattern holds if two classes in a grade exist and the new maximum of 56 (28 times 2) is exceeded by one. Only in this case, the addition of a third teacher reduces the average class size slightly less dramatically to 19. As the number of classes per grade increases, the cost of adding new students decreases and the variance in class sizes narrows.

Since 60 percent of Boston's 73 elementary schools average fewer than two teachers per grade, officials must either add an additional teacher to accommodate the one or two pupils who exceed the maximum or direct these students to other schools.[10] Boston's choice plan helps officials control this variance by allowing them to limit enrollment at a given school and assign students to one of their other top choices. But, analysis of each of Boston's 645 elementary classes shows that class sizes vary from 15 to 31. It is not unusual to find class size differences of eight or nine students from one grade to another in the same school. Comparing an average weighted by the number of students in each size of class to the unadjusted average class size helps to gauge the extent and frequency of class size variation. For example, if a class size average of 25 were composed of two classes each 25 in size, then the average would describe the experience of all students. However, an average of 25 also describes the combination of one class of 40 and one of 10. In this very different picture, where the weighted average totals 34, 80 percent of students spend their time in a class of 40. The weighted average regular education class size for Boston's elementary schools totals 23.2 as compared to the unweighted average of 22.2, a 5 percent difference. Because some variance in class size is unavoidable, it is hard to interpret, much of it could be controlled without comparison to other districts. Nevertheless, the difference represents resources that might be redirected more strategically at the school level.

Assignment of teachers to subject areas represents the second source of potential variation in regular class sizes. Analysis of class sizes by subject could highlight opportunities for shifting resources toward topics requiring more individualized attention. Boston's staffing formula leaves teacher allocation by subject to the schools' discretion. An analysis of a representative high school and middle school shows that at both schools, class sizes for the academic subjects of English, science, math, and history are consistently smaller than for other courses. At South Boston High, academic classes averaged 22 students as compared to the overall average of 25, with language arts classes lowest at 21, math the next lowest at 23, and history at 25 on average. Health, foreign language, and art classes averaged 30, 30, and 34 respectively. Similarly, at the Edison Middle School classes in English, math, science, and social studies averaged 18 while fine arts, computer science, home economics, gym, and foreign language averaged 26. Thus, while class sizes differed by subject these differences were consistently linked to an educational strategy.[11]

Implications and Alternatives

In the examples above, class sizes often differed not for educational reasons, but for management or union driven ones. A system that gave a school more options in assigning students to classes would reduce the non-educationally driven differences in class size. There are several ways to increase the number of options for placing students. One way would be to create schools with more classes per grade. This could be done by consolidating two small schools. Alternatively, reconfiguring the schools, for example, K-2 and 3-5, would allow the district to concentrate students at the same grade levels without creating larger schools. The above strategies assume that students must necessarily be grouped by age. Relaxing this assumption opens other opportunities to make the resources equal for students with similar educational needs. For example, a growing number of Boston schools like Trotter and Mason combine grades to even class sizes, organizing students by developmental level rather than age. This practice could give schools more flexibility in where they might assign students.

Grouping students in new ways could improve student achievement at the same time it helps to optimize the utilization of teaching resources. Recent research suggests that age-graded groupings in elementary schools for every subject may not effectively serve children's differing developmental stages (Slavin, 1991). A review of 64 research studies on the effects of nongraded schools by Barbara Pavan finds that nearly all studies show students in ungraded schools performing as well or better on standardized test measures than students in traditional schools. Further, these studies find that students in ungraded schools may have higher self-esteem and social skills (Pavan, 1992).

Barriers to Flexible Student Groupings

Reducing the rigid distinctions between classes of students that cause unintended variation in class sizes raises both contractual and educational issues. No rules prohibit combining grades except that the BTU contract specifies that classes which combine grades may not exceed 22 in size, a stipulation these proposals meet. The district can limit the number of very small special programs in two ways. First, the district can and does determine in advance where these programs should be housed and cap the number of students at the school level to avoid the addition of resources to cover the marginal student. Second, as described above, school professionals can reduce the numbers of such special programs by mainstreaming students and using any funds earmarked to help these students in a more inclusive setting.

But, as with reducing pull-out practices, changing the rationale for grouping students can embody a different approach to education. The new strategy would require teachers to work together to diagnose each child and determine how to best group them. Teachers could group students for part of the day based on reading level and reorganize them throughout the day. Alternatively, they could create more diverse multi-aged classrooms which used students as peer tutors. Either strategy would call on teachers to employ a sophisticated array of diagnostic tools and instructional strategies. Helping teachers develop these capabilities will take time and resources. But, in urban systems where students have such diverse needs, one might argue that teachers should employ these practices even without such a reorganization.

CURRICULUM AND SCHEDULING PRACTICES

Current Teacher Allocation

Curriculum and scheduling traditions constitute the final set of practices limiting time for individual attention and teacher planning. The majority of Boston's middle and high schools schedule students for seven, 45-minute periods a day. With five classes of 20 students in middle school and 25 in high school, each teacher works with 100 to 125 students per day. No matter how astute and dedicated the teacher, diagnosing the skill levels, learning capabilities, and personal needs of this number of students becomes close to impossible. For example, as Sizer (1985) vividly describes in *Horace's Compromise*, the idea of grading over 100 student papers with any kind of frequency, thoroughness, or timeliness, verges on the absurd.

Implications and Alternatives

Two factors drive teaching loads: the number of students per class and the number of classes taught. One way to reduce teaching loads would be to reduce average class size. But, without adopting the above strategies of reducing pull-out programs and changing p&d practices, this solution requires prohibitive increases in the number of teachers. The other way to lower teaching loads is to reduce the number of classes a teacher has. Doing this without dramatically increasing cost, demands rethinking curriculum and scheduling to lengthen the duration of classes with each teacher. This can be done either by combining traditionally separate subjects or by breaking the year into segments during which students study a small number of subjects intensively.

Boston's Timilty School integrates the subjects of English and history allowing lower teaching loads and more in-depth exploration of subjects. A national survey of restructuring efforts found a growing number of schools integrating subjects to reduce teacher loads and improve learning (Rettig & Canady, 1994). And, creating interdisciplinary courses is a key component of reaching targeted teacher loads of 80 in Sizer schools (Lusi & Watkins, 1989). A simplified example quantifies how integrating subjects could reduce student loads in a Boston middle school from 100 to 40 without changing the number of teachers. Combining the curriculum time spent on English, history and writing (which is taught as a specialty two or three times a week in many Boston schools) would create one class lasting two and one-half periods, or an hour and 52 minutes. In this scenario, each teacher would teach two classes of 20 each for a little less than two hours each day. The teachers would teach the same amount of time each day, but with 60 fewer students. Students would receive the same amount of weekly instruction in these subjects, but in an integrated course from the same teacher.

An alternative that retains subject specialties would be to have students study only one or two subjects at a time. So, in an overly simplistic example, the school year could be broken into five time periods. Students could take one major subject during each of these "quintmesters." Teachers could work for the equivalent of five periods with 20 students in their subject area. This plan would not reduce instructional time for students and would have the effect of reducing a teacher's load to 20 students. A variation on this theme might have students take three subjects at once which rotated over the year. Instead of having five, 45-minute classes of 20 students each all year, teachers would have three, 75-minute classes of 20 students each for a portion of the year. The student's total instructional time in that subject would remain unchanged, but the teacher's student load would drop from 100 to 60 throughout the year (Rettig & Canady, 1994). The Copernican Plan, in use at a growing number of schools, operates on this principle. As Joseph Carroll describes it, the schedule change is "not an end in itself but a means to create a classroom environment that fosters vastly improved relationships between teachers and students and that provides more manageable work loads for both teachers and students" (Carroll, 1994, p. 106).

As with the earlier example of integrating pull-out students, creating longer periods and moving to a more interdisciplinary approach are consistent with emerging research on how people learn best. Research by cognitive scientists such as Sylvia Scribner, Senta Raizen, and Lauren Resnick suggest that our standard approach of breaking competencies into small parts and teaching them separately out of context defies the way individuals learn. To develop expertise, they argue, students must be able to see the links between subjects and apply new concepts across disciplines (Raizen, 1989; Scribner, 1986). As Kathleen Cushman writes, an interdisciplinary approach "...centers more on rigorous habits of mind than on a fruitless quest for coverage...and then applies those habits to content across disciplinary lines" (Cushman, 1993, p. 1). Evaluations of the Copernican Plan, just one way to create lower teacher loads and longer class periods, found that the eight schools implementing it experienced "improvements of a magnitude seldom, if ever reported from a group of our nation's high schools" (Carroll, 1994, p. 112).

Barriers to Changing Curriculum Practices

Creating interdisciplinary courses and reorganizing the school day would require significant regulatory and contractual changes and also raises profound issues regarding both the practice of teaching and the education and certification of teachers. Massachusetts state law allows teachers to practice only in subjects in which they have been certified to teach, regardless of knowledge or skill. Even if a teacher did possess the required certification, the BTU contract does not allow a teacher to instruct more than one course outside her primary subject. Thus, unless all of the teachers in a school building agreed, a teacher would not be allowed to teach two interdisciplinary humanities courses at once.[12]

Once again, if a school did devise a restructuring plan that teachers agreed on, teachers would face the equally difficult task of creating new curriculum and classroom techniques to take advantage of it. Interdisciplinary courses that meet for longer time periods each day let teachers make more sophisticated links between subject areas and employ a wide range of materials and strategies in the classroom. Making these kinds of changes requires time and investment in training.

CONCLUSION

This analysis demonstrates the power of reexamining the use of all teaching resources at once. As many of the current patterns of teacher allocation have evolved as incremental responses to existing teaching conditions and traditions, it follows that small adjustments may not break this cycle. Four strategies reduce the specialization and fragmentation of teaching resources that collectively limit a school's flexibility to best match its teaching resources to student needs. These include:

- Increasing the portion of teachers who serve children in regular classroom settings while at the same time reducing the segregation of students with special needs.
- Considering new ways of providing for teacher planning and professional time and redeploying teachers now used to cover this time to reduce average teacher student ratios.
- Eliminating age grading or combining grades and programs within schools to reduce the unevenness in class sizes where the differences are not educationally driven.
- Rethinking school schedules to reduce the number of students for which each teacher has responsibility.

Together, these strategies have the potential to free 42 percent of Boston's teaching resources for alternative ways of organizing schools. Three measures provide an indication of the size of the opportunity for freeing teaching resources in any district:

- the difference between the district-wide pupil teacher ratio and the actual average class size for regular students,
- the difference between the regular class size average of and the weighted average size of these classes which incorporates the variance in class sizes and more accurately reflects students' experience,
- the actual average number of students secondary teachers have responsibility for each day (teacher loads).

These findings have implications for the burgeoning privatization and charter schools movements as well as for teachers unions. Organizations willing to start from scratch will have the opportunity to design more student-centered schools with dramatically lower class sizes and teacher loads for all using the same resources urban schools currently have. Because these changes require comprehensive change in a host of union regulations and assumptions, an incremental approach to bargaining may not work.

Finally, this analysis demonstrates that reconsideration of resources and educational strategy must proceed hand in hand. As Figure 4.6 summarizes, though regulatory and contractual constraints to rethinking teacher allocation exist, the most enduring and difficult challenge is changing how teachers and administrators work in schools. Designing new school organizations with more

Strategy	Example Alternatives	Obstacles
Reduce use of pull-out programs	-Integrate special education teachers and students with regular students where possible -Use Chapter 1 resources within regular program	-Teacher commitment and capability -District incentives -State Class size regulations -Parental consent
Redesign provision of teacher planning and development time	-Not covering p&d with additional, certified teachers -Extending/Restructuring teacher work day -Exceeding class size maximums for certain subjects and lessons	-Union/Mgmt Contract -State class size regulations
Modify formula driven student assignment to create more flexibility	-Combine grades -Limit numbers of students in special programs at school level	-District Policy -Teacher commitment and capability
Restructure daily schedule in secondary schools	-Create interdisciplinary course -Offer fewer subjects at a time and rotate throughout year	-Teacher Commitment and capability -Teacher certification regulations -Union/Mgmt contract

Figure 4.6. Opportunities for Freeing Teaching Resources

fluid lines between teachers, students, and subjects will require teachers to exercise broader professional discretion. This professional discretion would replace the rigid specialization of teachers by subject and type of student, the separation and segregation of students by special need and by age, and the lock-step daily schedule in secondary schools. Replacing these bureaucratic practices with teacher judgment could free significant resources. But, doing so demands short-term investment to help school professionals develop the knowledge and skills to realize this potential.

ACKNOWLEDGMENTS

I would like to thank the following people for their comments on earlier drafts of this article: Richard Elmore, Richard Murnane, and Susan Moore Johnson of Harvard University, and Willis Hawley of the University of Maryland. I would also like to thank Ellen Guiney who initiated this project from the Boston mayor's office and spent hours in interviews, finding and reviewing data and presenting findings throughout the community. Finally, I would like to thank the many people in the Boston Public School system for their generous assistance providing data and reviewing findings. Peter Rowe, then Deputy Superintendent of Management and Budget and Mary Grassa O'Neill, then Superintendent of the North Zone provided special help and support. The opinions expressed in this article are solely mine.

NOTES

1. NCES data show that school districts nationwide spent 62 percent on instruction in 1990-1991 (NCES, 1992, Table 8). Bruce Cooper attempted to refine this categorization in eight districts and also found that districts spent an average of 62 percent on instruction (Center for Workforce Preparation, 1993). Boston spending by function breaks out as follows.

Boston Public School Spending by Function
All Funds, 1990-1991*

	$ Millions	% of Budget
Teachers/Aides		
• Kindergarden	9.9	
• Regular Ed	140.9	
• Special Ed	86.2	
• Bilingual Ed	27.4	
• Vocational Ed	4.9	
Subtotal	269.3	61%
Instructional Supplies	4.7	1%
Student Support		
• Academic**	9.7	
• Special Ed Evaluation	7.2	
• Athletic	1.7	

• Social and Medical 7.9		
• Other (Police, monitors)......2.3		
Subtotal....................28.8		7%
School Administration......27.2		6%
*Central/Zone Admin****. 28.8		7%
Operations		
• Regular Transportation......20.2		
• Special Ed Transport..........10.7		
• Facilities & Maintenance....40.5		
• Non-instructional equip...... 6.4		
• Reserves...................................3.2		
Subtotal....................81.0		18%
GRAND TOTAL.................439.8		100%

* These totals do not include spending for school lunches.

** Academic Student Support includes librarians, library aides, testing supplies and services, community field coordinators and audio support services.

*** Central Admin includes central administration, as well professional and technical support, school committee staff and legal costs.

2. This analysis is based on the "Boston Public Schools Position Control Personnel Report. no. 2 FY 91" which lists each position sorted by fund, location of work, such as school, zone, or central office, title and function such as teacher, administrator, and so forth.

3. This analysis was conducted jointly with Richard Fossey and is summarized in Fossey and Miles (1992).

4. Those students with needs that cannot be served within the public school receive publicly supported private schooling. These private school teachers and students are not included in class size averages.

5. The systemwide ratio of students for every teacher in Boston is calculated by dividing the total number of students 48,346 (Table 4.2) by the total number of classroom teachers 3,673 (Table 4.3) which equals 13.16. These totals do not include kindergarten teachers or students and teachers in private placement or in BPS which are exclusively for special education or vocational education. The total number of classroom teachers includes only those teachers who work in schools for the majority of the day.

Calculating the actual class size for regular education requires adding both the regular education students and the special education mainstreamed students to get a total of all students who spend time most of their time in regular classrooms. So, systemwide this would be 30,281 plus 6,426 or 36,707. This number would then be divided by the number of teachers adjusted appropriately to account for the fact that elementary school teachers include specialists who teacher art, music, and gym and do not have their own classes. Similarly, secondary school teachers teach five of seven periods, or 70 percent of the day. Therefore, the total effective number of teachers was derived by adding the number of elementary school teachers with their own classrooms and taking 70 percent of the secondary school teachers.

In calculating alternative scenarios, the same adjustments to create the effective number of teachers are made. To determine the class sizes that would be possible by integrating Chapter 1 and resource teachers, these teachers are added into the divisor without changing the number of students. This is because Chapter 1 and resource students already spend the majority of their time in the regular classroom and are included in regular class size averages. Calculating possible class sizes with substantially separate and bilingual teachers requires adding both the effective number of teachers and the number of substantially separate students to the calculation since these students spend almost all of their time outside the classroom.

Table 4.2. Students By Type Sped Sped

	Reg Ed	*Sped Main*	*Sped Separate*	*Bil*	*Total*
Elem	12,742	2,886	1,906	3,826	21,360
Middle	5,581	1,891	1,276	1,396	10,144
High	11,958	1,649	1,284	1,951	16,842
Total	30,281	6,426	4,466	7,173	48,346

Source: Boston Public School Data 1/31/91.

Table 4.3. Teachers By Type

	Reg Ed	*Ch1*	*Sped Resource*	*Sped Separate*	*Bil*	*Total*
Elem*	768	145	141	293	264	1.611
Middle	520	59	75	146	93	893
High	788	49	75	134	123	1,169
Total	2,076	253	291	573	480	3,673

Note: *This number includes specialists who cover p&d time without them totals would be 701,132,129,268,241.

Source: Boston Public School Records 1/31/91.

6. Current class size maxima have the opposite effect. The BTU contract specifies lower maxima of 20 for occupational education classes like home economics and industrial arts, reasoning that they require close supervision of students using specialized equipment.

7. The percent of teachers freed by integrating teachers and students into the regular classroom is derived in two steps. First, the number of teachers required to cover planning and development are subtracted from the total, leaving only the teacher time which will be spent instructing classes. This separates the 142 elementary school specialists and 619 or 30 percent of the secondary teachers necessary to provide p&d. Second, I determined how many teachers would be needed to serve the bilingual and special education students now spending most of their time outside the regular classroom in class sizes averaging 22.7 and subtracted these teachers from the actual number of teachers currently serving these students. This process highlights the number of teachers who provide support beyond the ratio of one teacher for 22.7 which regular education students now receive. So, for example, bilingual teachers comprise 13 percent of BPS teachers. First, adjusting for planning and development time, bilingual classroom time uses 10.7 percent of all teaching resources. If all bilingual students returned to class sizes at the regular education average of 22.7 and the necessary bilingual teachers followed, only 76 bilingual teachers would have no classroom, or 2 percent of all teachers.

8. Most elementary schools have two or fewer first-grade classes averaging 21 in size. The addition of a third teacher would mean than 42 students would have three teachers, thus allowing class sizes of 14.

9. See American Arbitration agreement 1139-2311-90.

10. Analysis of end of year staffing reports by school for 1990-1991 shows that 26 percent of BPS' 73 elementary schools average less than one teacher per grade, 34 percent between one and two, 31 percent between two and three, and 85 have three or more teachers.

11. This analysis is based on staffing and scheduling data from two schools chosen based on their size, student composition, and staffing practices to be most representative of BPS middle and high schools. The data represent 1992-1993 staffing.

12. AAA case number 11e39-1942-89.

REFERENCES

Alexander, A.J., & Monk, D. (1987). *Attracting and compensating America's teachers.* Cambridge, MA: Ballinger.

Alexander, A.J. (1974). *Teachers, salaries and school district expenditures.* Santa Monica, CA: The Rand corporation.

Allington, R.L., & McGill-Franzen, A. (1989). Different programs: indifferent instruction. In D.K. Lipsky & A. Gartner (Eds.), *Beyond separate education: Quality education for all* (pp. 75-97). Baltimore, MD: Paul H. Brookes.

Barro, S.M. (1989). Fund distribution issues in school finance: priorities for the next round of research. *Educational Evaluation and Policy Analysis, 11*, 17-30.

Barro, S.M., & Carroll, S. (1975). *Budget allocation by school districts: An analysis of spending for teachers and other resources.* Santa Monica, CA: Rand Corporation.

Bird, T., & Little, J.W. (1986). How schools organize the teaching profession. *Elementary School Journal, 86*, 504.

Brimelow, P. (1986). Are we spending too much on education? *Forbes*, 26.

Brown, D.J. (1990). *Decentralization and school based management.* London: Falmer Press.

Burrup, P. et al. (1988). *Financing education in a climate of change* (4th ed.). Boston: Allyn and Bacon.

Canady, R.L., & Rettig, M.D. (1993). Unlocking the lockstep high school schedule. *Phi Delta Kappan, 70*, 310-314.

Carroll, J.M. (1994). Time for the Copernican plan. *Phi Delta Kappan, 76* (2), 104-114.

Center for Workforce Preparation. (1993). *Final report to the Lilly Endowment Co. on the quality education school finance project.* Washington, DC: U.S. Chamber of Commerce.

Common Destiny Alliance. (1992). *Realizing our nation's diversity as an opportunity: Alternatives to sorting America's children.* Nashville, TN.

Cooper, B.S., & Sarrel, R. (1992). Managing for school efficiency and effectiveness. *National Forum of Educational Administration and Supervision Journal, 8* (3), 3-38.

Cooper, B.S. (1993). School-site cost allocations: Testing a micro-financial model. Paper prepared for the American Education Finance Association.

Council of Urban Boards of Education. (1990). *Survey of public education in the nation's urban school districts.*

Cuban, L. (1989). The "at risk" label and the problem of urban school reform. *Phi Delta Kappan, 70* (10), 780-784.

Cushman, K. (1993). Creating an interdisciplinary curriculum. *Horace 9* (9).

David, J. (1983). School improvement and programs for special populations. In A. Odden (Ed.), *School finance and school improvement.* Cambridge, MA: Ballinger Press.

Donahoe, T. (1993). Finding the way: Structure, time and culture in school improvement. *Phi Delta Kappan, 72*, 298-305.

Ferguson, R.F. (1991). Pay for public education: New evidence on how and why money matters. *Harvard Journal on Legislation, 28*, 465-498.

Finn, C.E. (1993, September). Funding radical school reform. *Governing*, 14-15.

Folger, J. (Ed.). (1992). Evidence from project star about class size and student achievement. *Peabody Journal of Education, 67* (1), 17-33.

Fossey, R., & Miles, K. (1992). School based management in the Boston public schools: Why isn't it working? Paper for the Boston Mayor's Office.

Gartner, A., & Lipsky, D.K. (1987). Beyond special education: Toward a quality system for all students. *Harvard Educational Review, 57* (4), 367-390.

________. (1989). *The yoke of special education: How to break it.* Rochester, NY: National Center on Education and the Economy.

Glass, G.V., & Smith, M.L. (1979). Meta-analysis of research on class sizse and achievement. *Educational Evaluation and Policy Analysis, 1*, 2-26.

Guthrie, J.W. (1988). *Understanding school budgets.* Washington DC: U.S. Department of Education.

Hallahan, D.P., Keller, C.E., McKinney, J.D., Lloyd, J.W., & Bryan, T. (1988). Examining the research base of the regular education initiative: Efficacy studies and the adpative learning environments model. *Journal of Learning Disabilities, 21* (1), 29-35.

Hanushek, E.A. (1986). The economics of schooling. *Journal of Economic Literature, 24*, 1141-1177.

________ . (1989). The impact of differential expenditures on student performance. *Educational Researcher, 18* (4), 45-52.

________ . (1991). When school finance 'reform' may not be good policy. *Harvard Journal on Legislation, 28*, 423-456.

Harp, L. (1993, March 31). Dollars and sense: Reformers seek to rethink school financing to make it a powerful lever of change. *Education Week*, 9-14.

Hess, G.A., Jr. (1992, May). Reorienting a school district's funding priorities by state mandate. *Network News and Views*, 70-80.

Kakalik, J.S., Furry, W.S., Thomas, M.A., & Carney, M.F. (1981). *The cost of special education.* Santa Monica, CA: Rand Corporation.

Kazal-Thresher, D.M. (1992). Educational expenditures and school achievement when and how money can make a difference. *Educational Researcher, 22* (2), 30-32.

Kirst, M.W. (1988). Internal allocation of resources within U.S. school districts: Implications for policymakers and practitioners. In D.H. Monk & J. Underwood (Eds.), *MicroLevel school flnance: Issues and implications for policy* (pp. 365-388). Cambridge, MA: Ballinger.

Levin, H.M. (1987). *New schools for the disadvantaged.* Stanford, CA: Stanford University.

________ . (1988). *Accelerated schools for at-risk students.* Center for Policy Research in Education, No. RR-010. New Brunswick, NJ.

________ . (1991). *Building school capacity for effective teacher empowerment.* Center for Policy Research in Education, No. RR-019. New Brunswick, NJ.

Lohr, S. (1992, April 5). Lean and mean is foreign to schools. *The New York Times.*

Lusi, S.F., & Watkins, J.M. (1989). Mythos: Getting to 1:80. Coalition of Essential Schools, Brown University.

Marsh, D., & Sevilla, J. (1992). Middle school reform. In A.R. Odden & L. Picus (Eds.), *Rethinking school finance.* New York: McGraw Hill

Massachusetts Advocacy Center. (1990). *Locked in/locked out.* Boston.

McDonnell, L.M., & Pascal, A. (1988). *Teachers unions and educational reform.* Santa Monica, CA: Rand Corporation.

McDonnell, L.M. (1983). School improvement and fiscal retrenchment. In A. Odden (Ed.), *School finance and school improvement.* Cambridge, MA: Ballinger Publishing.

Moore, M.T., Strang, E.W., Schwartz, M., & Braddock, M. (1988). *Patterns in special education service delivery and cost.* Washington, DC: Decision Resources Corporation.

Murnane, R.J. (1991). Interpreting the evidence on 'does money matter?' *Harvard Journal on Legislation, 28*, 457-465.

National Center for Education Statistics. (1992). *The digest of education statistics.* Washington, DC: US Governmental Printing Office.

Nelson, F. (1983). School district response to labelling, cost and programmatic incentives in sped. *Journal of Education Finance, 3*, 80-98.

Novack, J. (1992, October 12). What do we get for our school dollars? *Forbes.*

Nye, B. (1992, May). Smaller classes really are better. *The American School Board Journal, 3*, 1-34.

Oakes, J. (1985). *Keeping track: How schools structure inequality.* New Haven, CT: Yale University Press.

Odden, A.R. (1989). The educational excellence movement: Politics and impact. *Economics of Education Review, 8* (4), 367-371.

________ . (Ed.). (1992). *Rethinking school finance: An agenda for the 1990s.* San Francisco: Jossey-Bass.

Odden, A.R., & Picus, L.O. (1992). *School finance: A policy perspecitve.* New York: McGraw-HIll.

Odden, A., & Webb, D. (Eds.). (1983). *School finance and school improvement: Linkages for the 1980s.* Cambridge,MA: Ballinger Publishing.

Pavan, B.N. (1992). The benefits of non-graded schools. *Educational Leadership, 50*(2), 22-25.

Picus, L.O. (1993). *Allocation and use of educational resources: School level evidence from the schools and staffing survey.* Los Angeles, CA: Center for Research in Education Finance. Working Paper No. 37.

Peltzman, S. (1993, July-August). Political factors in public school decline. *The American Enterprise.*

Poupard-Tice, D. (1988). *Developing an instructional grouping policy: A participatory model.* Ann Arbor, MI.

Raizen, S. (1989). *Reforming education for work: A cognitive science perspective.* National Center for the Reform of Vocational Education Policy Paper.

Raphael, E.S., Singer, J.D., & Walker, D.K. (1985) Per pupil expenditures on special education in three metropolitan school districts. *Journal of Education Finance, 11*, 69-88.

Retting, M.D., & Canady, R.L. (1993). Unlocking the lockstep high school schedule. *Phi Delta Kappan, 70*, 310-314.

Sarason, S.B. (1982). *The culture of school and the problem of change.* Boston: Allyn & Bacon.

Scribner, S. (1986). Thinking in action: Some characteristics of practical thought. In R.J. Sternberg & R.K. Wagner (Eds.), *Practical intelligence.* Cambridge, UK: Cambridge University Press.

Singer, J.D. (1988). Per pupil expenditures for special education: to whom are limited resources provided? U.S Department of Education, Final Report for Grant No. G008630147.

Sizer, T.R. (1992). *Horace's school.* Boston: Houghton Mifflin Co.

________ . (1995). *Horace's compromise.* Boston: Houghton Mifflin Co.

Slavin, R.E. (1990a). Ability grouping and student achievement in secondary schools: A best-evidence synthesis. *Review of Educational Research, 6*, 471-499.

________ . (1990b). On success for all: Defining 'success,' defining 'all.' *Remedial and Special Education, 11* (4), 60-61.

________ . (1991). Chapter 1: A vision for the next quarter century. *Phi Delta Kappan, April,* 586-592.

Slavin, R.E., Madden, N.A., Dolan, L.J., Wasik, B.A., Ross, S.M., & Smith, L.J. (1994, April). Whenever and wherever we choose: The replication of 'success for all.' *Phil Delta Kappan,* 639-647.

Skrtic, T.M. (1991). The special education paradox: equity as the way to excellence. *Harvard Educational Review, 61* (2), 148-206.

Tyack, D., & Tobin, W. (1994). The 'grammar' of schooling: Why has it been so hard to change? *American Educational Research Journal, 31* (3), 453-479.

Watts, G.D., & Castle, S. (1993). The time dilemma in school restructuring. *Phi Delta Kappan, 70*, 306-310.

Wechsler, D. (1990, June 25) Parkinson's law 101. *Forbes*, 52-56.

Wheelock, A. (1992). *Crossing the tracks: How "untracking" can save America's schools.* New York: The New Press.

Wohlstetter, P., & Buffett, T.M. (1992). Promoting school based management: Are dollars decentralized too? In A.R. Odden & L.O. Picus (Eds.), *Rethinking school finance: An agenda for the 1990s.* New York: McGraw Hill.

PART II

THE BATTLE FOR FINANCIAL EQUITY: FEDERAL, STATE, LOCAL DISTRICT, AND SCHOOL SITE

Chapter 5

MEETING THE CHALLENGE OF DEVOLUTION:

HOW CHANGING DEMOGRAPHY AND FISCAL CONTEXTS AFFECT STATE INVESTMENTS IN EDUCATION

Martin E. Orland and Carol E. Cohen

INTRODUCTION

As the direction of public policy points unambiguously to a larger state and local leadership role in delivering and paying for public services, the near- and long-term fiscal outlook for states takes on added relevance to those interested in education finance. The impending devolution of program responsibility and authority from Washington to states and localities means that these governments will be increasingly expected to design and fund strategies for

Advances in Educational Productivity, Volume 6, pages 101-125.

ISBN: 0-7623-0082-5

meeting the needs of their citizens. What financing challenges is the education sector likely to face in light of this changed context? What policy implications are suggested by this financing outlook?

Unlike most other children's services, education revenue is derived almost exclusively from state and local sources. The overall proportion of federal financial support for elementary and secondary education is currently under 7 percent So, at first glance, it might seem as though a smaller federal role would not have much impact on education service provision. However, on closer inspection, it is clear that federal devolution can be expected to affect education financing in at least two ways. First, states and school districts will be increasingly called on to provide financing to support the special categories of funding in which federal financing currently plays a major role. For instance, disadvantaged students, students with disabilities, and the limited English proficient have for years been particularly dependent on targeted federal assistance programs. And more recently, through programs such as Goals 2000 and the School-to-Work Opportunities Act, the federal government has begun to provide states and school districts with the marginal resources needed to stimulate broad system-wide education reforms grounded in the principle of all students achieving at dramatically higher levels.

Perhaps even more significantly, federal devolution will cause increased competition for the general state and local education dollar. As federal for programs like welfare and Medicaid are reduced, additional pressures will be placed on state and local budgets to accommodate the shortfalls. Because education comprises such a large share of state and local budgets (comprising about 38 cents out of every state and local tax dollar in 1992), it may become an especially inviting revenue target.

To what extent will the education sector in states and local communities be equipped to meet the financial challenges brought on by a declining federal role? Answering this question satisfactorily requires that we appreciate the factors most directly associated with patterns and trends in education spending. Even a cursory look at average per pupil expenditure levels reveals how much spending was over three times higher in New Jersey than in Utah. Even more dramatically, while average per pupil education spending rose $110 per year between 1970 and 1989, it increased at a rate of only $15 per year between 1990 and 1994.

The purpose of this analysis is to better understand what drives spending contrasts like these, and what this portends for future spending on education. It is our hope that such an understanding will enable policymakers at all levels to make more informed decisions on public educational investments.

ANALYTIC APPROACH AND FRAMEWORK

This study analyzes the fiscal challenges ahead for states in financing education by examining patterns of state spending for these services and the major factors influencing them. In addition to examining recent cross-sectional state data on per pupil education spending, we look at spending changes between 1970, 1980, and 1992. Our approach is based on the assumption that the factors and relationships that are significant in explaining current and recent state spending will continue to affect such spending in the future.

The hypothesis framing our analysis is that three broad factors can influence state spending for education and other children's services.

- the *need* (or the demographic capacity) of states to provide education services;
- the *ability to pay* (or the fiscal capacity) of states to provide such services; and
- the *willingness to pay* (or the fiscal effort) of states and localities in support of these services.

The relevance of each of these factors to the level of per pupil education spending in each state is highlighted below.

Service Needs (Demographic Capacity)

The magnitude of states' overall need for educational services can have a major impact on the amount of resources they invest in each pupil. In this study, we measure this factor—which we have labeled its demographic capacity—by determining a state's *ratio of population to pupils.*[1] All else being equal, states who are able to spread their educational costs across a larger population base (i.e., those with higher population to pupil ratios), can more easily generate a given level of per pupil spending than can states with greater numbers of students relative to their population.

Ability to Pay (Fiscal Capacity)

The ability to pay—or fiscal capacity—of a state can also have an impact on per pupil education spending. A state's fiscal capacity represents the potential of that state to generate resources for public purposes. Thus, the higher the level of a state's fiscal capacity, the greater is its presumed ability to fund all public services, including education. Likewise, the stronger the growth of fiscal capacity, the greater is a state's ability to increase spending for those services.

As with indicators of need, there are many possible choices for indicators of state fiscal capacity. Some, such as per capita income, are based on broad measures of economic activity within a state, while others, such as the Representative Tax System developed by the Advisory Commission on Intergovernmental Relations, focus more directly on the revenue-raising potential of state and local governments in a state. And some measure are better at capturing the potential of states to "export" taxes to, or raise revenues from, nonresidents than are others. Nevertheless, the fiscal capacity indexes for most states tend to differ very little depending on what measure is used—except in those states with relatively large oil production or tourism industries, where the potential for tax exporting is the greatest.

In this report, we use *per capita personal income* as the indicator of a state's ability to pay for educational services. Per capita income is a major component of a state's capacity to raise revenues for public services because most taxes are paid from the income of a state's residents. Per capita income is the most widely used indicator of fiscal capacity and the most readily available for the years examined in this study.[2]

Willingness to Pay (Fiscal Effort)

The third major factor that can affect state per pupil education spending is a state's willingness to pay for education. Willingness to pay is captured by the "fiscal effort" a state makes. Fiscal effort relates a state's actual revenues or spending to its fiscal capacity. Because fiscal capacity varies across states, a state with lower fiscal capacity will have to use a greater share of its capacity to achieve the same service levels as a state with higher fiscal capacity (all else being equal) and vice versa. Fiscal effort thus provides a measure of the relative burden placed on a state's resources, or the "effort" made to achieve the service levels that are provided.[3]

Fiscal effort can be measured for the total of all revenues or spending (i.e., the overall fiscal effort of a state) or for selected categories. In this study, we use education spending per $100 of personal income as our measure of fiscal effort for education. Because we use personal income (on a per capita basis) as our indicator of fiscal capacity, we also use it in defining our measure of fiscal effort.

Relationship of Service Needs, Ability to Pay, and Willingness to Pay

We have noted that service needs, ability to pay (or fiscal capacity), and willingness to pay (or fiscal effort) can each independently affect state per pupil spending levels. But how do these factors interrelate in each state to affect spending? The interaction for education spending can be described by the following mathematical identity developed by Gold and his colleagues (1996, p. 26).

$$\text{School Spending/Pupils} = \text{Spending/Income} * \text{Income/Population} * \text{Population/Pupils}$$

In this equation, we see that education service levels (school spending per pupil) are a multiplicative function of fiscal effort (spending in relation to personal income); fiscal capacity (per capita income); and demographic capacity (the ratio of population to pupils). The mathematical identity, in effect, decomposes state per pupil spending levels into demographic capacity, fiscal capacity, and fiscal effort components. As will be seen later in this study, by relating each component in a state to its corresponding value for the United State as a whole, its relative contribution in explaining that state's spending can be observed.

FEATURES OF OUR PRESENTATION AND DATA

Our analysis primarily examines national patterns and trends in the data and discusses what they are likely to mean for most states in the future. Because of the great variation among state, however, we also present state-by-state data and highlight significant variations among states or regions where they exist.[4]

Our work relies on data compiled by Gold and his colleagues (1995), as well as an analysis prepared for The Finance Project by the same authors (1996). That database contains state-by-state data as well as national data on state spending for education (as well as other children's services) and related economic and demographic factors. Some of the features of that database are described below.

Definitions of State Education Spending

State spending on education is defined as all current spending for public elementary and secondary education in a state. It thus excludes expenditures for capital improvement and focuses only on ongoing expenditures for K-12 education. Education spending includes funds from federal revenue sources. However, because the federal contribution to state elementary-secondary education spending is relatively small (in 1992, less than 7%),[5] this measure of education spending primarily reflects the commitment of states—including their local governments—to education spending from their own resources.

Time Period

The database includes education data for 1992—the most recent year for which all the data were available—as well as historical data for 1970, 1980, and 1992.

Adjustments for Inflation and Differences in Price Levels

Inflation reduces the value of a dollar of spending over time. To adjust for this effect, all fiscal data are presented in constant 1992 dollars.[6] Our comparisons of revenue, spending, and income data over time thus represent real changes in the levels of these variables, after accounting for the effects of inflation.

Likewise, differences in price levels among locations can bias interstate comparisons because of their effect on the purchasing power of families and governments. A family with a $40,000 annual income in Boston, for example, has much less purchasing power that one with the same income living in Jackson, Mississippi. However, because valid and reliable state-level price-adjusted data are less readily available over time than nonadjusted data, most of the data in this report are unadjusted. In those instances where we have used an existing index to adjust for interstate price-level differences (see, for example, Table 5.1), the results suggest that such adjustments narrow but do not eliminate the wide variations among states.[7]

KEY FINDINGS: PATTERNS AND TRENDS IN EDUCATION SPENDING

Elementary and secondary education constitutes by far the largest single category of spending by state and local governments. In 1992 states devoted 34 percent of their tax revenues to finance K-12 education, compared with about 20 percent for health, 12 percent for higher education, and 8 percent for social welfare. Thirty-eight cents of every state and local tax dollar that year supported education (Gold et al., 1996, tables 2-14).

The magnitude of state and local educational investments should not obscure the fact that states vary considerably in both their levels of education spending and rates of expenditure growth. As noted earlier, New Jersey spend over $9,000 per pupil in 1992, a figure that is about three times greater than that for Utah. Even when spending is adjusted for differences in the cost of living, substantial differences remain (Table 5.1). And while the last two decades were periods of substantial real growth in educational expenditures overall (Table 5.2), spending disparities among the states have remained relatively constant (Table 5.3).

In this section we attempt to document factors that influence education spending. Specifically, we address the question of how strongly indicators of demographic capacity (educational need), fiscal capacity (ability to pay), and fiscal effort (willingness to pay) can explain spending patterns and trends among the states. To the extent that any of these factors appear salient, we can use this knowledge to make more informed judgments regarding the prospects for education spending in the future.

Table 5.1. Education Spending per Pupil Rank Ordered by State, 1992

State	*Unadjusted Spending*	*Index (U.S. = 100)*	*Spending Adjusted for Cost Differences**	*Index (U.S. = 100)*
United States	$ 5,421	100	$ 5,421	100
New Jersey	9,317	172	7,302	135
New York	8,527	157	7,251	134
Alaska	8,450	156	6,387	118
Connecticut	8,017	148	6,258	115
Vermont	6,944	128	6,855	126
Maryland	6,679	123	5,808	107
Pennsylvania	6,613	122	6,186	114
Rhode Island	6,546	121	6,017	111
Massachusetts	6,408	118	5,344	99
Michigan	6,268	116	6,725	124
Wisconsin	6,139	113	6,658	123
Delaware	6,093	112	5,544	102
Oregon	5,913	109	6,231	115
Wyoming	5,812	107	6,144	113
New Hampshire	5,790	107	5,341	99
Ohio	5,694	105	6,116	113
Illinois	5,670	105	5,870	108
Maine	5,652	104	5,618	104
Montana	5,423	100	5,901	109
Hawaii	5,420	100	4,091	75
Minnesota	5,409	100	5,760	106
Washington	5,271	97	5,368	99
Nebraska	5,263	97	5,835	108
Florida	5,243	97	5,687	105
Colorado	5,172	95	5,219	96
West Virginia	5,109	94	5,872	108
Iowa	5,096	94	5,669	105
Indiana	5,074	94	5,600	103
Kansas	5,007	92	5,594	103
Nevada	4,926	91	5,027	93
Virginia	4,880	90	5,304	98
Missouri	4,830	89	5,279	97
California	4,746	88	4,280	79
Kentucky	4,719	87	5,356	99
Texas	4,632	85	5,147	95
North Dakota	4,441	82	4,979	92
South Carolina	4,436	82	4,979	92
Arizona	4,381	81	4,554	84
Georgia	4,375	81	4,850	89
Louisiana	4,354	80	4,937	91
South Dakota	4,173	77	4,699	87
Oklahoma	4,078	75	4,618	85

(continued)

Table 5.1. (Continued)

Arkansas	$4,031	74	$4,602	85
New Mexico	3,765	69	4,088	75
Tennessee	3,692	68	4,148	77
Alabama	3,616	67	4,100	76
Idaho	3,556	66	3,891	72
Mississippi	3,245	60	3,738	69
Utah	3,040	56	3,304	61
Ratio between Highest- and Lowest-Spending States	3.1 to 1		2.2 to 1	
50-State Average	5.330		5,384	
Standard Deviation	1.323		889	
Coefficient of Variation	0.25		0.17	

Notes: * *Spending* adjusted by cost index prepared by F. Howard Nelson, American Federation of Teachers.

Source: Gold et al., 1995.

Table 5.2. Percentage Change in Real per Pupil Spending, by Region: 1970-1980, 1980-1992, and 1970-1992

Region	*State*	*1970-1980*	*1980-1992*	*1970-1992*
United States		26.6	37.0	73.4
New England	Connecticut	15.7	90.2	120.0
	Maine	19.9	77.9	113.2
	Massachusetts	49.2	30.5	94.7
	New Hampshire	20.5	73.5	109.0
	Rhode Island	32.7	44.5	91.8
	Vermont	12.5	99.6	124.6
Mid-Atlantic	Delaware	44.5	22.2	76.7
	Maryland	28.7	47.6	89.9
	New Jersey	42.8	67.6	139.3
	New York	18.6	41.4	67.7
	Pennsylvania	30.7	49.7	95.7
Great Lakes	Illinois	29.4	25.8	62.8
	Indiana	17.5	54.8	81.9
	Michigan	32.8	36.3	81.0
	Ohio	29.2	57.5	103.6
	Wisconsin	27.6	42.3	81.5
Plains	Iowa	25.3	25.8	57.6
	Kansas	28.2	32.3	69.5
	Minnesota	20.1	30.1	56.2
	Missouri	24.2	43.2	77.8
	Nebraska	32.8	40.5	86.6
	North Dakota	26.5	32.8	68.0
	South Dakota	25.7	25.5	57.8

(continued)

Table 5.2. *(Continued)*

Region	*State*	*1970-1980*	*1980-1992*	*1970-1992*
Southeast	Alabama	34.7%	28.8%	73.5%
	Arkansas	26.0	47.0	85.2
	Florida	17.3	59.3	86.9
	Georgia	25.7	54.5	94.2
	Kentucky	41.9	59.2	126.0
	Lousiana	25.7	39.5	75.4
	Mississippi	51.0	11.9	69.1
	North Carolina	30.3	49.1	94.3
	South Carolina	30.0	45.3	88.9
	Tennessee	31.3	29.6	70.2
	Virginia	26.5	42.2	79.9
	West Virginia	30.3	52.7	99.0
Southwest	Arizona	24.5	27.6	58.8
	New Mexico	30.8	6.2	39.0
	Oklahoma	45.0	21.5	76.2
	Texas	39.6	38.8	93.7
Rocky Mountain	Colorado	49.2	22.6	82.9
	Idaho	25.1	23.0	53.9
	Montana	44.0	25.7	81.0
	Utah	20.4	5.3	26.7
	Wyoming	34.2	32.0	77.2
Far West	Alaska	91.4	2.6	196.4
	California	18.9	20.1	42.9
	Hawaii	25.5	34.0	68.2
	Nevada	23.5	35.4	67.2
	Oregon	32.3	26.1	66.8
	Washington	27.6	17.8	50.4

Note: Number of pupils is average daily attendance.

Source: Gold et al., 1995.

Table 5.3. Real Education Spending per Pupil, by Region: 1970-1980, and 1992 (in 1992 constant dollars)

Region	*State*	*1970*	*1980*	*1992*
New England	Connecticut	$3,644	$4,216	$8,017
	Maine	2,651	3,178	5,652
	Massachusetts	3,291	4,911	6,408
	New Hampshire	2,770	3,338	5,790
	Rhode Island	3,414	4,531	6,546
	Vermont	3,902	3,479	6,944
Mid-Atlantic	Delaware	3,448	4,984	6,093
	Maryland	3,517	4,526	6,679

(continued)

Table 5.3. (Continued)

	New Jersey	$3,893	$5,559	$9,317
	New York	5,084	6,031	8,527
	Pennsylvania	3,379	4,416	6,613
Great Lakes	Illinois	3,483	4,507	5,670
	Indiana	2,789	3,279	5,074
	Michigan	3,464	4,599	6,268
	Ohio	2,797	3,615	5,694
	Wisconsin	3,383	4,315	6,139
Plains	Iowa	3,234	4,052	5,096
	Kansas	2,954	3,279	5,007
	Minnesota	3,464	4,159	5,409
	Missouri	2,716	3,373	4,830
	Nebraska	2,820	3,746	5,263
	North Dakota	2,644	3,345	4,441
	South Dakota	2,644	3,324	4,173
Southeast	Alabama	2,084	2,808	3,616
	Arkansas	2,176	2,742	4,031
	Florida	2,805	3,291	5,243
	Georgia	2,253	2,831	4,375
	Kentucky	2,088	2,963	4,719
	Lousiana	2,483	3,122	4,354
	Mississippi	1,920	2,899	3,245
	North Carolina	2,345	3,056	4,555
	South Carolina	2,349	3,052	4,436
	Tennessee	2,169	2,848	3,692
	Virginia	2,713	3,432	4,880
	West Virginia	2,567	3,345	5,109
Southwest	Arizona	2,759	3,434	4,381
	New Mexico	2,709	3,544	3,765
	Oklahoma	2,314	3,355	4,078
	Texas	2,391	3,338	4,632
Rocky Mountain	Colorado	2,828	4,218	5,172
	Idaho	2,310	2,890	3,556
	Montana	2,996	4,314	5,423
	Utah	2,398	2,887	3,040
	Wyoming	3,280	4,402	5,812
Far West	Alaska	4,303	8,237	8,450
	California	3,322	3,951	4,746
	Hawaii	3,222	4,045	5,420
	Nevada	2,946	3,638	4,926
	Oregon	3,544	4,690	5,913
	Washington	3,506	4,474	5,271
50-State Average		2,947	3,862	5,330
Range		3,165	5,495	6,277
Standard Deviation		605	971	1.323
Coefficient of Variation		0.21	0.25	0.25

Note: Number of pupils is average daily attendance. Figures were adjusted by the State and Local Government Implicit Price Deflator from the Economic Report of the President (1992=100).

Source: Gold et al., 1995.

Table 5.4. Ratio of Population to Public School Enrollment, by Region: 1970, 1980, and 1992

Region	*State*	*1970*	*1980*	*1992*	*1992 Rank*
United States		4.42	5.39	6.00	
New England	Connecticut	4.64	5.47	6.70	7
	Maine	4.13	4.94	5.71	28
	Massachusetts	4.92	5.55	7.08	1
	New Hampshire	4.76	5.35	6.26	11
	Rhode Island	5.15	6.19	7.06	2
	Vermont	4.37	5.12	5.85	24
Mid-Atlantic	Delaware	4.14	5.76	6.66	8
	Maryland	4.34	5.43	6.61	9
	New Jersey	4.88	5.73	7.00	4
	New York	5.26	5.94	6.83	6
	Pennsylvania	5.00	6.03	7.06	3
Great Lakes	Illinois	4.75	5.59	6.24	12
	Indiana	4.20	5.05	5.86	22
	Michigan	4.11	4.97	5.88	21
	Ohio	4.34	5.33	6.13	15
	Wisconsin	4.47	5.44	6.07	18
Plains	Iowa	4.25	5.32	5.68	32
	Kansas	4.31	5.55	5.59	37
	Minnesota	4.11	5.19	5.72	27
	Missouri	4.31	5.60	6.12	16
	Nebraska	4.45	5.44	5.69	30
	North Dakota	4.20	5.54	5.35	40
	South Dakota	4.01	5.15	5.34	42
Southeast	Alabama	4.16	5.13	5.66	34
	Arkansas	4.16	5.01	5.41	38
	Florida	4.72	6.28	6.87	5
	Georgia	4.09	5.00	5.63	35
	Kentucky	4.54	5.38	5.75	26
	Lousiana	4.33	5.17	5.34	41
	Mississippi	3.96	4.94	5.68	33
	North Carolina	4.24	5.04	6.15	14
	South Carolina	3.96	4.94	5.68	33
	Tennessee	4.37	5.23	5.94	20
	Virginia	4.29	5.16	6.19	13
	West Virginia	4.35	5.00	5.62	36
Southwest	Arizona	4.15	5.18	5.70	29
	New Mexico	3.66	4.65	5.01	45
	Oklahoma	4.14	5.09	5.39	39
	Texas	4.01	4.83	4.98	46
Rocky Mountain	Colorado	4.02	5.18	5.68	31
	Idaho	3.93	4.60	4.60	48

(continued)

Table 5.4. *(Continued)*

	Montana	3.46	4.25	3.87	50
	Utah	3.46	4.25	3.87	50
	Wyoming	3.81	4.74	4.49	49
Far West	Alaska	3.85	4.50	4.79	47
	California	4.29	5.65	5.95	19
	Hawaii	4.16	5.63	6.49	10
	Nevada	3.88	5.18	6.10	17
	Oregon	4.31	5.52	5.85	23
	Washington	4.07	5.22	5.77	25

Note: The 1992 population-to-enrollment ratios in this table differ slightly from the population-to-pupil ratios in Table 5.8. Because of data constraints, full enrollment data were used to create the ratios in this table, while average daily attendance data were used to create the ratios in Table 5.8.

Source: Gold et al., 1995.

We begin this discussion by examining data on demographic capacity, fiscal capacity, and fiscal effort between 1970 and 1992. We then relate these factors to changes nationally in per pupil spending level over this period, as well as to differences among the states in per pupil spending. Finally, we discuss the implications of these findings for future education spending.

Education Spending and Demographic Capacity: The Importance of the Relative Size of the School Population

Our indicator of the need for educational services in a state (or its demographic capacity) is the size of the total population relative to the number of children enrolled in the public schools. As noted in the previous section, the higher a state's ratio of population to pupils, the higher its demographic capacity, and vice versa. By this measure, demographic capacity to support per pupil education spending increased by 26 percent between 1970 and 1992. Growth occurred in every state and was especially pronounced during the decade of the 1970s. States with the highest demographic capacity tend to be overwhelmingly in the Northeast and Great Lakes regions, while those with the lowest demographic capacity are consistently found among the Southwest and Rocky Mountain states (Table 5.4).

As suggested earlier, higher demographic capacity should be good news for per pupil education spending. This is because the financial burden of educating children who are in the public education system can be spread among more taxpayers. Conversely, having more pupils relative to the population at large would be expected to make it more difficult to generate high per pupil spending levels. Simple correlations appear to support this hypothesis. In 1992 the correlation between our measure of education need and per pupil education spending was -0.51, while the correlation between percentage enrollment growth and per pupil spending growth (between 1970 and 1992) was -0.26.

Table 5.5. Real Per Capita Income, by Region: 1969, 1979, and 1991

		Per Capita Income (in 1991 dollars)			Percentage Change		
Region	*State*	*1969*	*1979*	*1991*	*1969-1979*	*1979-1991*	*1969-1991*
United States		$12,636	$16,485	$19,199	30.5	16.5	51.9
New England	Connecticut	15,960	19,619	25.844	22.9	31.7	61.9
	Maine	10,370	13,394	17,330	29.2	29.4	67.1
	Massachusetts	14,075	17,273	22,796	22.7	32.0	62.0
	New Hampshire	12,398	15,880	20,961	28.1	32.0	69.1
	Rhode Island	12,640	15,476	19,451	22.4	25.7	53.9
	Vermont	11,178	14,133	17,811	26.4	26.0	59.3
Mid-Atlantic	Delaware	14,648	16,809	20,317	14.8	20.9	38.7
	Maryland	13,780	17,670	22,483	28.2	27.2	63.2
	New Jersey	14,923	18,890	24,744	26.6	31.0	65.8
	New York	15,225	17,820	22,925	17.0	28.6	50.6
	Pennsylvania	12,570	16,498	19,638	31.2	19.0	56.2
Great Lakes	Illinois	14,390	18,425	20,622	28.0	11.9	43.3
	Indiana	12,189	15,776	17,275	29.4	9.5	41.7
	Michigan	13,445	17,427	18,693	29.6	7.3	39.0
	Ohio	12,872	16,343	18,001	27.0	10.1	39.8
	Wisconsin	12,179	16,416	17,970	34.8	9.5	47.5
Plains	Iowa	11,901	16,169	17,102	35.9	5.8	43.7
	Kansas	11,636	16,571	18,259	42.4	10.2	56.9
	Minnesota	12,365	16,683	19,289	34.9	15.6	56.0
	Missouri	11,715	15,712	18,105	34.1	15.2	54.5
	Nebraska	11,725	15,661	15,594	48.6	6.2	57.9
	North Dakota	9,876	14,677	15,594	48.6	6.2	57.9
	South Dakota	9,747	14,206	16,419	45.7	15.6	68.5

(continued)

Table 5.5. (Continued)

Southeast	Alabama	$9,024	$12,814	$15,601	42.0%	21.7%	72.9%
	Arkansas	8,617	12,637	14,458	46.7	14.4	67.8
	Florida	12.007	15,857	19,203	32.1	21.1	59.9
	Georgia	10,439	13,920	17,636	33.3	26.1	59.9
	Kentucky	9,700	13,522	15,442	39.4	14.2	59.2
	Louisiana	9,521	14,010	15,442	39.4	14.2	59.2
	Mississippi	7,841	11,644	13,210	48.5	13.4	68.5
	North Carolina	9,962	13,293	16,810	33.4	26.5	68.7
	South Carolina	9,216	12,509	15,469	35.7	23.7	67.8
	Tennessee	9,720	13,447	16,489	38,3	22.6	69.6
	Virginia	11,669	15,959	20,074	36.8	25.8	72.0
	West Virginia	9,167	13,299	14,665	45.1	10.3	60.0
Soutwest	Arizona	11,390	15,235	16,760	33.8	10.0	47.1
	New Mexico	9,571	13,542	14,818	41.5	9.4	54.8
	Oklahoma	10,482	15,083	15,656	43.9	3.8	49.4
	Texas	11,129	16,118	17,440	44.8	8.2	56.7
Rocky Mountain	Colorado	12,182	17,191	19,745	41.1	14.9	62.1
	Idaho	10,578	14,014	15,854	32.5	13.1	49.9
	Montana	10,631	14,558	15,793	36.9	8..5	48.5
	Utah	10,022	13,452	14,737	34.2	9.5	47.1
	Wyoming	11,682	18,612	18,295	59.3	-1.7	56.6
Far West	Alaska	15,371	22,665	21,592	47.5	-4.7	40.5
	California	14,887	19,241	20,880	29.2	8.5	40.3
	Hawaii	14,698	17,486	21,621	19.0	23.7	47.1
	Nevada	14,840	19,263	20,774	29.8	7.8	40.0
	Oregon	12,100	16,723	17,789	38.2	6.4	47.0
	Washington	13,475	17,890	20,163	32.8	12,7	49.6

Source: U.S. Department of Commerce, Bureau of Economic Analysis, State Summary Tables (August 1994) (SA1-3, SA51-52), 1929-1993. 1948-1993 and calculations by The Finance Project.

Education Spending and Fiscal Capacity: The Importance of the Size of the Revenue Pie

While education service needs were declining over the past two decades, the capacity to finance these services was growing at a healthy rate. As measured by changes in real per capita income, state fiscal capacity grew by 31 percent in the 1970s, and 17 percent between 1979 and 1991 (Table 5.5). State growth patterns were generally consistent with national trends during both decades, with the notable exception of most states in the New England and Mid-Atlantic ares, where gains were stronger during the 1979-1991 period than from 1969 to 1979. Most high-capacity states can be found in the New England and Mid-Atlantic regions, while low-capacity states tend to predominate in the Southeast, Southwest, and Rocky Mountain areas.[8]

Obviously, all things being equal, fiscally strong states can support generous per pupil education spending levels more easily than can states with a poorer resource base. We would thus expect to see both higher levels of per pupil spending in states with greater fiscal capacity, and also higher rates of per pupil spending growth among states where fiscal capacity gains where greatest. Simple correlations reveal a strong relationship between per capita income and per pupil spending for 1992 (0.80) and a weaker, but still substantial relationship between 1970-1992 changes in per capita income and changes in per pupil spending (.52).

Education Spending and Fiscal Effort: The Importance of Educational Resource Commitments

As noted earlier, the degree to which a state taps its available resource capacity—or its fiscal effort—is a third factor explaining per pupil education spending. A state devoting a larger share of its available resources for education will spend more per pupil than a comparable state (in terms of both demographic and fiscal capacity) making a more modest resource commitment.

Educational effort is really a function of two components. One is the *size* of the government sector in the state relative to overall available resources. A larger government revenue base means more resources potentially available to support educational expenditures.

The second critical component of educational effort is the *share* of government resources supporting education. Differences among states in the education share of the government pie, as well as changes in that share over time, can profoundly affect education effort levels and, ultimately, per pupil expenditures.

Nationally, educational effort remained relatively stable from 1970 through 1992. It declined a bit in the 1970s, before growing modestly from 1980 to 1992 (Table 5.6). Analyzing educational effort by its two core components reveals that

Table 5.6. Growth in Education Spending in Relation to Personal Income and Education's Share of Tax Revenue: 1970, 1980, and 1992

	1970	*1980*	*1992*
Current Education Spending per $100 of Personal Income	$4.46	$4.30	$4.36
State-Local Education Revenue as a Percentage of Total State-Local Tax Revenue	43.5	39.0	38.2

Source: Gold et al., 1995.

the small overall decrease in education effort is entirely attributable to smaller education shares of state and local tax bases. General state and local government tax effort levels remained relatively unchanged from 1970 to 1992. But the share of this resource base going to education declined from approximately 44 percent to 38 percent (Table 5.6). Most of this decline occurred in the 1970s and was a function of reduced local (rather than state) government education revenue shares.

The overall stability in educational effort should not obscure significant changes in some states in recent years (Table 5.7). Massachusetts, for example, increased its effort by nearly 30 percent in the 1970s, only to decreasing it by about 25 percent during the 1980s and early 1990s. Wyoming did the opposite, decreasing effort significantly in the 1970s (18%) and increasing it by an even greater rate (33%) between 1980 and 1992. Over the entire 1970-1992 period, 14 states witnessed double-digit decreases in educational effort, while seven experienced double-digit gains. Significantly, educational effort is not strongly associated with region. States with high and low effort levels, and with small and large recent changes in their relative resource commitments to education, can be found in all parts of the country.

Examining Interstate Spending Differences

As just discussed, differences in levels of demographic capacity, fiscal capacity, and fiscal effort can each help to explain varied levels of per pupil education spending among the states. But are there patterns in the relative influence of these factors that can be useful in projecting the challenges ahead in education spending? Using the identity developed by Gold, we examine the relative contribution of each in determining per pupil expenditure levels.

Table 5.8 arrays states by their 1992 per pupil spending levels, alongside measures of demographic capacity (the ratio of population to number of pupils), fiscal capacity (per capital income), and education effort (education spending per $100 of personal income). The data are indexed to the national average to facilitate comparisons. A few things are noteworthy about

Table 5.7. Education Spending per $100 of Personal Income, by Region: 1970, 1980, and 1992

Region	State	Spending per $100 Income			Percentage Change		
		1970	1980	1992	1970-1980	1980-1992	1970-1993
United States		$4.46	$4.30	$4.36	-3.6	1.5	-2.2
New England	Connecticut	4.07	3.69	4.31	-9.4	16.8	5.8
	Maine	5.02	4.68	5.23	-6.9	11.9	4.2
	Massachusetts	3.78	4.86	3.68	28.5	-24.2	-2.6
	New Hampshire	3.74	3.73	4.00	-0.3	7.1	6.8
	Rhode Island	4.09	4.47	4.44	9.3	-0.8	8.4
	Vermont	5.35	4.86	6.24	-9.3	28.6	16.6
Mid-Atlantic	Delaware	4.56	4.89	4.14	7.3	-15.4	-9.2
	Maryland	4.49	4.37	3.99	-2.7	-8.7	-11.1
	New Jersey	4.21	4.78	5.02	13.6	5.2	19.4
	New York	4.94	5.10	4.78	3.1	-6.2	-3.3
	Pennsylvania	4.29	4.28	4.42	-0.4	3.3	2.9
Great Lakes	Illinois	3.96	3.98	3.89	0.6	-2.2	-1.7
	Indiana	4.28	3.92	4.69	-8.4	19.8	9.7
	Michigan	5.05	5.27	5.23	4.2	-0.8	3.4
	Ohio	4.00	3.97	4.63	-0.6	16.6	15.9
	Wisconsin	4.83	4.56	5.17	-5.7	13.5	7.1
Plains	Iowa	5.23	4.60	4.94	-12.1	7.3	-5.6
	Kansas	4.62	3.90	4.46	-15.5	14.3	-3.4
	Minnesota	5.57	4.85	4.61	-13.0	-4.9	-17.3
	Missouri	3.91	3.58	3.87	-8.5	8.0	-1.2
	Nebraska	4.44	4.34	4.81	-2.3	10.9	8.4
	North Dakota	5.29	4.36	4.97	-17.5	14.0	-6.0
	South Dakota	5.57	4.45	4.50	-20.1	1.0	-19.2

(continued)

Table 5.7. *(Continued)*

Southeast	Alabama	$4.51	$4.23	$3.86	-6.3%	-8.6%	-14.4%
	Arkansas	4.73	4.25	4.83	-10.0	13.6	2.2
	Florida	3.99	3.37	3.65	-15.7	8.5	-8.5
	Georgia	4.18	3.92	4.11	-6.3	5.0	-1.6
	Kentucky	3.77	3.91	4.72	3.7	20.7	25.2
	Lousiana	4.84	4.11	4.99	-15.1	21.3	3.0
	Mississippi	5.00	4.73	4.49	-5.4	-5.2	-10.3
	North Carolina	4.47	4.46	4.11	-0.3	-7.9	-8.1
	South Carolina	5.15	4.72	4.66	-8.2	-1.4	-9.5
	Tennessee	4.14	3.96	3.50	-4.4	-11.5	-15.4
	Virginia	4.34	4.05	3.96	-6.7	-2.2	-8.8
	West Virginia	5.16	4.81	5.74	-6.9	19.2	11.0
Southwest	Arizona	4.72	4.32	4.14	-8.5	-4.1	-12.3
	New Mexico	6.29	5.43	5.29	-13.6	-2.7	-16.0
	Oklahoma	4.23	4.31	4.58	-12.5	0.1	-12.4
	Texas	4.09	4.08	4.86	-0.3	19.1	18.7
Rocky Mountain	Colorado	4.64	4.64	4.14	0.1	-10.8	-10.7
	Idaho	4.57	4.39	4.62	-3.9	5.3	1.2
	Montana	5.71	5.70	6.01	-0.2	5.5	5.2
	Utah	5.68	4.97	4.98	-12.5	0.1	-12.4
	Wyoming	6.00	4.91	6.52	-18.1	32.6	8.6
Far West	Alaska	5.93	7.64	7.59	28.9	-0.7	28.0
	California	4.33	3.75	3.73	-13.4	-0.4	13.8
	Hawaii	4.29	3.87	3.60	-9.7	-6.9	16.0
	Nevada	4.06	3.50	3.60	-13.9	2.9	-11.4
	Oregon	5.36	4.78	5.06	-10.9	5.9	-5.7
	Washington	5.15	4.65	4.21	-9.7	-9.4	-18.2

Source: Gold et al., 1995.

Table 5.8. Determinants of Education Spending per Pupil Relative to the U.S. Average, by Region: 1992

Region	*State*	*Education spending per pupil*	*Index (U.S.=100)*	*Ratio of pop. to pupils*	*Index (U.S.=100)*	*Per capita income*	*Index (U.S.=100)*	*Education spending per $100 pers. inc.*	*Index (U.S.=100)*
New England	Connecticut	$8,017	148	7.19	111	$25,844	135	$4.31	99
	Maine	5,652	104	6.23	96	17,330	90	5.23	120
	Massachusetts	6,408	118	7.63	118	22,796	119	3.68	85
	New Hampshire	5,790	107	6.92	107	20,961	109	4.00	92
	Rhode Island	6,546	121	7.59	117	19,451	101	4.44	102
	Vermont	6,944	128	6.25	97	17,811	93	6.24	143
Mid-Atlantic	Delaware	6,093	112	7.25	112	20,317	106	4.14	95
	Maryland	6,679	123	7.44	115	22,483	117	3.99	92
	New Jersey	9,317	172	7.50	116	24,744	129	5.02	115
	New York	8,527	157	7.79	121	22,925	119	4.78	110
	Pennsylvania	6,613	122	7.62	118	19,638	102	4.42	101
Great Lakes	Illinois	5,670	105	7.07	109	20,622	107	3.89	89
	Indiana	5,074	94	6.26	97	17,275	90	4.69	108
	Michigan	6,268	116	6.42	99	18,693	97	5.23	120
	Ohio	5,694	105	6.83	106	18,001	94	4.63	106
	Wisconsin	6,139	113	6.61	102	17,970	94	5.17	119
Plains	Iowa	5,096	94	6.04	93	17,102	89	4.94	113
	Kansas	5,007	92	6.15	95	18,259	95	4.46	102
	Minnesota	5,409	100	6.08	94	19,289	100	4.61	106
	Missouri	4,830	89	6.89	107	18,105	94	3.87	89
	Nebraska	5,263	97	6.06	94	18,047	94	4.81	110
	North Dakota	4,441	82	5.72	89	15,594	81	4.97	114
	South Dakota	4,173	77	5.65	87	16,419	86	4.50	103

(continued)

Table 5.8. (Continued)

Southeast	Alabama	$3,616	67	5.95	92	$15,601	81	$3.86	89
	Arkansas	4,031	74	5.77	89	14,458	75	4.83	111
	Florida	5,243	97	7.47	116	19,203	100	3.65	84
	Georgia	4,375	81	6.03	93	17,636	92	4.11	94
	Kentucky	4,719	87	6.47	100	15,442	80	4.72	108
	Louisiana	4,354	80	5.79	90	15,067	78	4.99	114
	Mississippi	3,245	60	5.48	85	13,210	69	4.49	103
	North Carolina	4,555	84	6.60	102	16,810	88	4.11	94
	South Carolina	4,436	82	6.16	95	15,469	81	4.66	107
	Tennessee	3,692	68	6.39	99	16,489	86	3.50	80
	Virginia	4,880	90	6.14	95	20,074	105	3.96	91
	West Virginia	5,109	94	6.07	94	14,665	76	5.74	132
Soutwest	Arizona	4,381	81	6.31	98	16,760	87	4.14	95
	New Mexico	3,765	69	4.80	74	14,818	77	5.29	121
	Oklahoma	4,078	75	5.69	88	15,656	82	4.58	105
	Texas	4,632	85	5.46	85	17,440	91	4.86	111
Rocky Mountain	Colorado	5,172	95	6.33	98	19,745	103	4.14	95
	Idaho	3,556	66	4.86	75	15,854	83	4.62	106
	Montana	5,423	100	5.71	88	15,793	82	6.01	138
	Utah	3,040	56	4.14	64	14,737	77	4.98	114
	Wyoming	5,812	107	4.88	76	18,295	95	6.52	149
Far West	Alaska	8,450	156	5.16	80	21,592	112	7.59	174
	California	4,746	88	6.09	94	20,880	109	3.73	86
	Hawaii	5,420	100	6.95	108	21,621	109	3.73	83
	Nevada	4,926	91	6.59	102	20,774	108	3.60	83
	Oregon	5,913	109	6.57	102	17,789	93	5.06	116
	Washington	5,271	97	6.21	96	20,163	105	4.21	97

Note: The population-to-pupil ratios in this table differ slightly from the 1992 population-to-enrollment ratios in Table 5.4. Because of data constraints—primarily the need to use a consistent measure of pupils in the two varibles of spending per pupil and the ratio of population to pupils for this identity—average daily attendance data were used to create the population-to-pupil ratios in this table, while fall enrollment data were used to create the ratios in Table 5.4.

Source: Gold et al., 1995.

these findings. First of all, as Gold (1996, p. 26), points out, there are few common patterns among the highest spending states; and different factors are associated with high education expenditures in different places. In Connecticut, bountiful tax capacity is the primary story (35% above the national average). Educational effort levels here are only about average. Vermont's high spending is completely attributable to its unusually high education effort rates; it devotes nearly half of its tax revenues to education spending, the fifth highest rate in the country (Gold, 1996, tables 2-12). By contrast, high incomes and favorable population/pupil ratios allow neighboring Massachusetts to spend generously with education effort levels that are only 85 percent of the national average.

The picture is much simpler in the lowest-spending states. As Gold also notes, virtually every one of the lowest-spending states has both low per capita income and low demographic capacity (p. 27). Significantly, most of the low-spending states (located primarily in the South) are making at least average levels of educational effort. However, the combination of a weak fiscal resource base and the need to support relatively large numbers of children result in low per-child service level.

To summarize, high state per pupil spending seems related to relatively unique combinations of demographic capacity, fiscal capacity, and fiscal effort. On the other hand, low spending is consistently explained in terms of low demographic and/or fiscal capacity.

Implications for the Future

The salience of income and demographic factors in explaining per pupil spending patterns and trends has two significant implications for education spending over the next decade. First of all, it is clear that more modest income growth and sharply increasing enrollments in most states mean that they will not witness growth rates in per pupil education spending that were commonplace during the 1970s and 1980s. As illustrated in Figure 5.1, school enrollment is projected to increase substantially through 2005, in sharp contrast to the declines that occurred in the 1970s and early 1980s. At the same time, growth in per capita income is projected through 2005 to be slower than the growth trends the nation has experienced in each half-decade since 1970, with the exception of the 1990-1994 period. It appears that recent income and enrollment changes have already contributed to dramatic slowdowns in per pupil spending growth during the early 1990s. The future outlook for these variables likely will further dampen per pupil spending gains through the year 2005. Further, these projections do not take into account the looming cutbacks in federal aid outside the education arena. These are likely to exacerbated fiscal pressures on the education sector as competition for available state and local dollars becomes more intense. It may, in fact, be difficult for overall levels of educational effort to remain stable in the coming decade in the face of such competition.

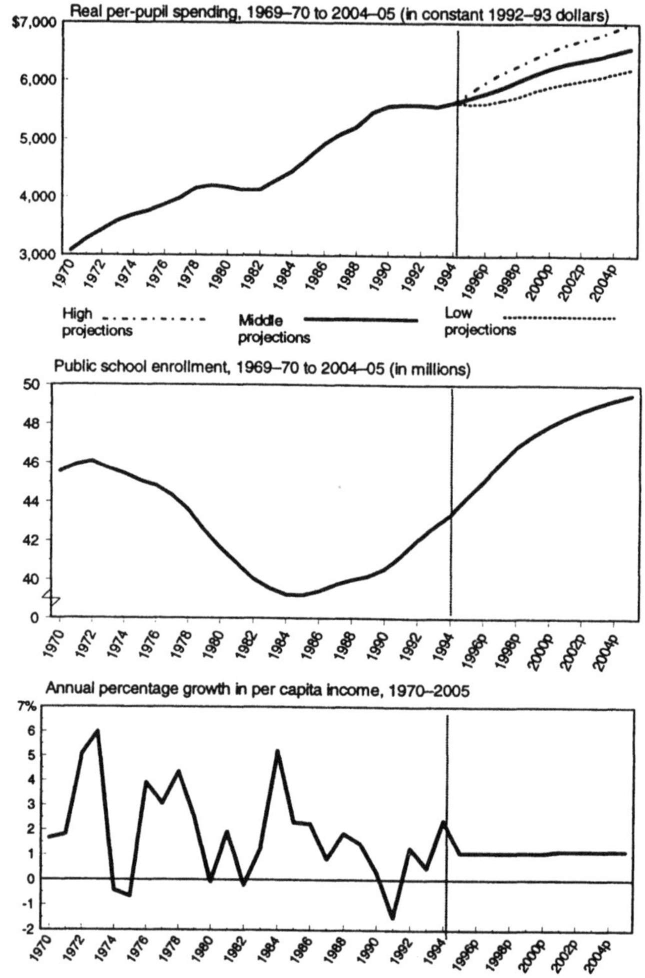

Change in real per-pupil spending (in constant 1992–93 dollars)

1970–74	$612
1975–79	$439
1980–84	$271
1985–89	$781
1990–94	$61
1995–99p	?
2000–05p	?

Change in public school enrollment (in thousands)

1970–74	-106
1975–79	-2,522
1980–84	-2,399
1985–89	981
1990–94	2,810
1995–99p	3,176
2000–05p	1,579

5-year average rates of growth in per capita income

1970–74	3.1%
1975–79	2.6%
1980–84	1.6%
1985–89	1.7%
1990–94	0.6%
1995–99p	1.1%
2000–05p	1.1%

Note: p = projected.

Source: Real per pupil spending and public school enrollment—U.S. Department of Education, National Center for Education Statistics. *Digest of Education Statistics, 1994*, and *Projections of Education Statistics to 2005*; per capita income growth—U.S. Department of Commerce, Bureau of Economic Analysis; historical data and projections for 2000 and 2005 contained in *Survey of Current Business*, July 1995; and calculations by The Finance Project.

Figure 5.1. Recent and Projected Trends in Spending Enrollment, and Income, 1970-2005

Second, it is extremely unlikely that disparities among states in their spending levels will be reduced much in the years ahead. Projections for income growth in the lowest-spending states are comparable, by and large, to those of the highest-spending states (U.S. Department of Education, 1995a, table D). And while a few of the lowest-spending states like Louisiana, Missippi, and North and South Dakota are projected to experience relatively small increases in enrollment growth, they are also among the states that are most dependent on federal aid (Gold, 1996, tables 1-9), and thus most likely to be affected by grant-in-aid cutbacks. Of the 10 highest-spending states, only three (Alaska, Maryland, and New Jersey) have higher-than-average projected enrollment growth through 2005. (U.S. Department of Education, 1995b, table 26).

SUMMARY AND CONCLUSIONS

Our primary purpose in studying state spending patterns and trends in education is to permit us to make more informed judgments about future state financial investments in this area. We employed a three-factor model consisting of measure of need (demographic capacity), ability to pay (fiscal capacity), and willingness to pay (fiscal effort), attempting to associate each of these with state per pupil spending levels for education.

Two of the three—need and ability to pay—demonstrated a stable, strong relationship to state education spending per pupil. States with larger proportions of children needing services (the school-age population) were consistently less likely to spend as much on a per-child basis as states with lower levels of overall need. Similarly, high-income states spent considerably more per child on education than did poor states. By contrast, the association between fiscal effort and state spending was relatively weak. Many low-spending states made large resource commitments to education relative to their income, while several high-spending states made a more modest education effort. It is the strong and consistent relationship between need, ability to pay, and state education spending which, when placed in the context of demographic and economic projections lead to concerns about future state educational spending. Factors associated with substantially higher spending levels in most states—in particular, healthy per capita income growth and declining school enrollments—are not likely to be present over the coming decade. Public school enrollments are expected to rise at a rate of about 1 percent per year between now and 2005 (U.S. Department of Education, 1995b, table 46). If more modest per capita income growth projections prove accurate, most states will be extremely hard pressed to maintain their spending patterns of recent years on education.[9]

The effects of cubacks in federal aid to states and localities in order to achieve a balanced federal budget can be expected to exacerbate further the fiscal stresses on states stemming from less favorable demographic and economic conditions. Federal aid currently augments state and local tax revenues by about one-third (Gold, 1996, tables 1-8). Thus, major reductions will undoubtedly put additional pressure on states to make up shortfalls in a variety of areas (e.g., transportation, higher education, and community development) by raising their own spending levels. Because tax hikes are unlikely and economic growth is expected to be modest, the revenue to fund any such increases may well come from reallocating existing resources. The education sector may become especially vulnerable, since it is the largest single functional component of state and local budgets, comprising 38 cents of every state and local tax dollar in 1992.

In conclusion, the salience of indicators of need and ability to pay in explaining state investment levels in education leads to the conclusion that most states will be greatly challenged in the years ahead in financing these programs. The extent of the financing challenge will vary considerably by state, but, in general, a combination of increased needs, slower growth in fiscal capacity, and reduced federal aid will make it extremely unlikely that the vast majority of states can sustain the per pupil spending pattens of the previous two decades.

ACKNOWLEDGMENTS

This paper is adapted from Martin E. Orland and Carol E. Cohen, *State Investments in Education and Other Children's Services: The Fiscal Challenges Ahead*, prepared for The Finance Project, Washington, D.C., September 1995. Martin E. Orland's views are his own, and no official support by the U.S. Department of Education is intended or should be inferred.

NOTES

1. This ratio is an admittedly incomplete proxy of state service needs or demographic capacity, but one that appears well-suited to the aggregate state-level analysis appearing in this paper. More refined measures would be expected to include other student population characteristics such as numbers from low-income families, numbers with limited English profienciency, numbers with disabilities and so on.
2. This report focuses on states' spending for education from their own resources. As such, the concept of fiscal capacity used in this study does not include federal aid. Athough, as noted earlier, federal grants to states can certainly be expected to affect their ability to finance educational services. Since major reductions in the size of these grants are likely, these impacts will be discussed separately from the influence of state fiscal capacity.
3. Fiscal effort may be a flawed proxy for willingnes to pay if there are external constraints on spending level, such as federal mandates, court orders, or state constitutional requirements for alloating funding.
4. Instrastate variations in spending and other variables can be as large or larger than interstate variations. Discussion of the extent and significance of intrastate variations is beyond the scope of this report, however.

5. Although the federal contribution to education spending ranged as high as 17.7 percent in one state (Mississippi), in 40 of the 50 states, the federal contribution was less than 10 percent.

6. Data on education spending were adjusted using the implicit price deflator for state and local government purchases. Per capita income data were adjusted using the fixed-weight personal consumption expenditure deflator.

7. There have been at least two recent efforts to develop indexes of state price-level differences. The index used in this report to illustrate the effect of adjusting state education spending for price-level differences was developed by F. Howard Nelson of the American Federation of Teachers Research Department and is contained in Nelson (1991). Another index used in this report to adjsut state fiscal capacity was developed by Herman Leonard and Monica Fryar (1994).

8. These regional patterns remain mostly intact after adjusting fiscal capacity for interstate differences in the cost of living. The effect of adjusting for cost-of-living differences is to reduce the variation in fiscal capacity among states. For example, after adjustment the overall fiscal capacity index of the New England region falls from 117 to 107, while that of the Southeast region rises from 89 to 97 (Gold et al., 1996, tables 1-5).

9. Some have pointed out that in addition to expected lower income growth rates, the increasing percentage of economic activity not subject to taxation will further handicap future state and local revenue-raising capacities (see Wallace, 1995).

REFERENCES

Gold, S.D. et al. (1995, September). *State investments in education and other children's services: Fiscal profiles of the 50 states.* Washington, DC: The Finance Project.

________. 1996, January. *How funding of programs for children varies among the 50 states.* Albany, NY: Center for the Study of the States.

Leonard., H., & Fryar, M. (1994). *By choice or by change* Boston: The Pioneer Institute.

Nelson, F.H. (1991, Spring). An interstate cost of living index. *Educational Evaluations and Policy Analysis, 13*, 103-111.

U.S. Department of Commerce. (1995, July). *Bureau of economic analysis.* Survey of Current Business. Washington, DC: U.S. Printing Office.

U.S. Department of Education (1995). National Center for Education Statistics. *Projcetions of education statistics to 2005.* NCES 95-169. Washington, DC: Government Printing Office.

Wallace, S. (1995). *The effects of economic and demographic changes on state and local budgets.* Washington, DC: The Finance Project.

Chapter 6

THE ASSESSMENT OF EQUAL EDUCATIONAL OPPORTUNITY: METHODOLOGICAL ADVANCES AND MULTIPLE STATE ANALYSES

Deborah A. Verstegen

INTRODUCTION

Following the historic *Serrano* decision (see Introduction), most of the school finance reforms implemented during the 1970s and early 1980s throughout the United States were designed to reduce the linkage between local wealth and revenue available to finance education; and to decrease disparities in per pupil funding between more and less affluent jurisdictions (Geske & LaCost, 1990). Beginning in 1983, with the onset of the education reform movement, issues related to the efficiency and productivity of the education system achieved

Advances in Educational Productivity, Volume 6, pages 127-148.

ISBN: 0-7623-0082-5

prominence. Overall the focus has been on linkages between U.S. economic growth, income inequalities and educational attainment in a knowledge society and global economy, strategies to transform the educational system to world-class status aimed at ambitious outcomes and enhanced productivity for all children and at all schools, and assessment systems to guide decision making and monitor progress over time. In this context, equity without excellence is no longer the goal.

This chapter discusses advances in the assessment of equal educational opportunity and efficiency directed at the question: How best might equal opportunities for enhanced productivity in schooling be achieved? Data from eight states and a 50-state survey inform this discussion. First the conceptual basis for the analysis of equal opportunity is discussed, then advances in the methodology are reviewed and research findings presented. Because the object of analysis in an assessment of equal opportunity is also used in efficiency research (along with measures of ability to pay) advances in assessment are relevant to both undertakings.

EQUAL EDUCATIONAL OPPORTUNITY

The concept of an equal educational opportunity is a widely held and deeply enshrined ideal of the American system of government. As early as 1813 Thomas Jefferson, in a letter to John Adams on the purposes of government, underscored the importance of a "natural aristocracy" to the fledgling nation. The "natural aristocracy" was intended to be built on talent and virtue unlike the "artificial aristocracy" that inhabited Europe at the time, which was founded on "wealth and birth." According to Jefferson, government's role in the realization of this ideal was to scale back discriminatory barriers to fair competition and equal opportunity: wealth and privilege. In this, the provision of education—particularly an equal educational opportunity—was critical.

Changes Over Time

Originally, the notion of an equal educational opportunity was taken to mean access to schooling; later this was expanded to mean access to a minimum educational program. Currently the concept of equal educational opportunity is being transformed in the context of the global economy and knowledge society to mean access to a quality, not a basic, educational program for all children and at all schools (Verstegen, 1996). In this, the issue of adequacy—or the sufficiency of the funding system to support a quality educational program—is conjoined with the issue of equity—that is, the fairness of the distribution of educational benefits and burdens for society.

Equity: Horizontal and Vertical

Generally three basic principles are utilized to define equal opportunity: (1) horizontal equity, (2) vertical equity, and (3) wealth neutrality (see Berne & Stiefel, 1984; Verstegen, 1990a). *Horizontal equity*, sometimes referred to as *arithmetical equality*, assumes that where different treatment cannot be justified, individuals clearly should be treated alike. This concept is affirmed by the principle of "one person, one vote." It indicates that no justifiable differences exist between individuals that would support differential franchise; therefore they should be treated the same.

In school finance the concept of equal treatment of equals, or horizontal equity, assumes equal dollars per pupil (pupil equity) or equal funding for equal tax rates (taxpayer equity). However, this does not mean that absolute equality in spending is required, due to the second principle that is part and parcel of the definition of equal opportunity: vertical equity.

Vertical equity refers to treatment of people in different circumstances according to their justifiable and relevant differences, or "unequal treatment of unequals." For example, capacity to pay is taken as a justifiable ground for individual differentiation under a system of progressive taxation. Variation in the local ability to pay for schooling is considered a justifiable basis for distinguishing among localities in the provision of state aid under a school finance equalization system.

Another justifiable ground for differentiation in the treatment of individuals is based on special needs. Need criteria presuppose some standard condition that a person or group would fall short of if the need were not satisfied. In school finance special needs criterion are the basis on which additional resources are allocated; they provide supplementary assistance for schools to mount programs and services to address special student needs that are not being met under the general education program. If left unaddressed these needs could impose a hardship on both the student and society. Children at risk of failing or dropping out of school, children with limited English proficiency (LEP), and children with disabilities comprise three groups of students with justifiable and educationally relevant special needs. As such, extra resources are not only justified but also required to meet these children's needs if equity is to be achieved. As Thurow (1985, p. 120) explains, although equity is often mistaken as a synonym for equality, equity does not arise when everyone is treated equally regardless of circumstances, but when everyone is treated fairly.

Equal Opportunity and Wealth Neutrality

Another principle used to assess fiscal equity is that of equal opportunity. It affirms that there should not be discrimination in the allocation of resources. In education finance this means there should be no relationship between per pupil revenues (expenditures) and certain characteristics such as race, alienage, sex, or wealth (a locality's ability to pay for education through local sources).[1] When revenue and wealth are unrelated, for example, the distribution is said to be "wealth neutral," and the "quality of a child's education is not a function of wealth other than the wealth of the state as a whole." However, as Berne explains: The goal of equal opportunity is both outcome equity and input equity; this remains a dual focus (1994, p. xii).

Adequacy

A concept that has recently emerged in school finance litigation and that complements notions of both horizontal and vertical equity, is *adequacy* (Verstegen & Whitney, 1995). It relates to the sufficiency of a distributional plan to meet its basic goals. That equity rests on an adequate distribution of resources is clear, in that the equality of an inadequate level of resources impoverishes the poor and the rich alike. Thus, school finance schemes must rest on principles of equity, equal opportunity, and adequacy if fairness and justice are to be realized.

METHODOLOGICAL CONSIDERATIONS—ASSESSING EQUITY, EQUAL OPPORTUNITY, AND ADEQUACY

Although there is no single technology for the measurement of adequacy in school finance, there exists a strong knowledge base for the measurement of equity, which has its genesis in the work *The Conceptualization and Measurement of Equity in School Finance* (Berne & Stiefel, 1984). However, recent advances in technology and methodological improvements—including the development and dissemination of systems that make school-level data more readily available (Cooper, 1994; Speakman et al., 1996), new equity measures or variations (Verstegen, 1996), alternative approaches for investigating relationships between factors that attempt to explain funding variations (cf. Fortune & O'Neil, 1994; Verstegen, 1994), and the emergence of empirically derived factors for vertical equity analysis (discussed herein)—have resulted in numerous issues. These are discussed below, along with more general questions related to the analysis of equity in school finance.

What Level of the System Should be Analyzed?

One of the first questions that must be answered in doing an equity analysis is: What level of the system should be analyzed? The answer to this depends on the question being investigated, data availability, and the purposes at hand. Researchers who are interested in examining questions related to the equity of schooling across the states employ an interstate equity analysis, which can provide information to guide federal policy aimed at equity and efficiency issues (Verstegen, 1992, 1994; Hertert, Busch, & Odden, 1994). On the other hand, an analysis of intrastate fiscal inequalities typically is directed to questions concerning features of the state finance system that may influence the equity and/or the constitutionality of the plan (cf. Verstegen, 1987; Verstegen & Salmon, 1989, 1991). Finally, researchers interested in district-level decisions concerning resource allocations and efficiency/productivity concerns, often focus on the equity of funding among schools, or on inequalities among classrooms within a school (Summers & Wolfe, 1967; Cooper 1994; for a discussion, see Berne & Stiefel, 1993). These analyses have been rarer due to the paucity of school-level finance data, but recently these data have been made more accessible (Cooper 1994; Speakman et al., 1996). Importantly, at all levels of the system three aspects of the analysis should be similar: the pupil unit of observation, the adjustments made to the revenue (expenditure) variable, and the equity statistics utilized.

The Unit of Analysis—Pupil or District

The pupil unit of analysis, based on a district's (state's/school's) average daily membership (ADM) or enrollment, is preferred in an analysis of fiscal equity in accordance with research and best practice (cf. Berne & Stiefel, 1984, chap. 1). This allows districts that have more pupils to be taken into account more heavily in the statistics than when a single figure is utilized and allows each child to receive an equal weight. The district unit of analysis, in effect, treats each district as if it has only one pupil and implies that each pupil in the larger district has a relatively smaller influence on the equity assessment than each pupil in smaller district. Using the pupil as the legitimate methodological unit of analysis, all statistics are computed with special procedures employed to accommodate weighted statistical measures for grouped data. This procedure treats these data as a distribution of pupils and assigns district (school) averages as attributes to each pupil. In the computation of the McLoone index, for example, the median pupil (not the median district) is utilized; for deciles (or quintiles), intervals are determined based on the percentage of pupils (not districts).

Objects of Analysis—Inputs, Outputs, Outcomes

Theoretically the possible objects of analysis could include the measurements of inputs, outputs, or outcomes. Because inputs provide the basis for an analysis of opportunities available to students, they are the preferred object. Multiple resource inputs can be compared—such as the distribution of teacher characteristics, class sizes, technology, and curriculum—but funding is the chief variable of interest because it allows localities choices in the mix, level, and intensity of physical and human resources that can be procured. Because the purpose of analysis is to measure differences in the opportunities for children and youth in schools and classrooms, all funding reaching students and supporting their programs and services must be included (see discussion to follow).

The Revenue (Expenditure) Variable and Adjustments

Revenues (or expenditures) have been the "standard object many people want distributed equitably among school children" (Odden, Berne & Stiefel, 1979, p. 8 ff.). Clune explains: dollars are "the 'master input' which purchase all other inputs" (1992, p. 737). However, the selection of dollars as the input variable raises several questions: Should all revenue—federal, state, and local—be considered in assessing equity? Should current operating expenditures be utilized (total state and local expenditures excluding debt service and capital outlay)? Should instructional expenditures be examined?

Because federal aid provides assistance to meet the excess needs of special populations—a legitimate basis for additional funds—it is excluded from the analysis of equity, the focus of which is unjustifiable disparities in revenue inputs (Odden, Berne, & Stiefel, 1979). Likewise, revenues designated for debt service and capital outlay are excluded as they are considered investment spending and due to yearly fluctuations cannot be readily attributed to the cost of educating a single pupil in a particular year. Co-curricular/enterprise funds and activity funds/foundations are not considered in the analysis due to the difficulty in capturing these revenue streams. The exclusion of these funds, except federal, likely results in the underestimation of both costs and disparities across objects.

The most commonly used variable in the analysis of equity is current operating expenditures—but adjustments to this variable to correct for excess costs are required before statistical computations can be executed, when these additional funds provide for legitimate variations in pupil needs or in the cost of education among districts (schools).[2] According to researchers, additional funding for pupil needs or district (school) costs should not be considered in a disparity analysis (cf. Berne & Stiefel, 1984).

As discussed earlier, equity for special student populations requires "differential expenditure levels. Unequal dollars are required to produce equal educational opportunities" (Hartman & Haber, 1981, p. 27). Likewise, unequal dollars are required to accommodate differences in the cost of providing educational services of similar quality among districts (schools) due to differences in, for example, economies of scale and/or in the variation in the purchasing power of the dollar. These are referred to as cost adjustments.

The judiciary (cf. *Serrano* v. *Priest*, 1986) concurs that adjustments are needed. For example, in *Serrano III* (1986), the deliberation of the court centered around two issues: (1) What unit of measure should the court use to measure equity? (2) How should equity be measured? (in Ward, 1990). The actual controversy centered on the extent to which various state categorical funding programs should be included in the analysis. The court decided that the measure to be equalized was general school spending per pupil exclusive of all categorical funding, reasoning that,

> The base revenue limit is the fundamental element of school funding that all school divisions receive. It is the *only* ongoing element of school finance that does not expressly relate to the varying special needs of either divisions or pupils. It is the common denominator (226 Cal.Rptr. 611, emphasis in the original).

The court also concluded that no single equity measure would suffice and held that multiple measures of equity would be used with general education funding (the base revenue limit) as the unit of measure and the pupil as the unit of analysis (Ward, 1990). Likewise, an analysis of equity should use multiple measures weighted by students; and adjustments should be made to correct for legitimate variations in revenue.

To adjust for funds associated with pupil needs and district costs, two lines of modification can be undertaken. First, legitimate excess costs for districts (schools) and students with special needs can be excluded in the revenue variable. Deducting these accounts is the approach most commonly utilized (when such adjustments are made at all) because research-based cost differentials have been unavailable; also these funds have been provided by the state and federal governments through categorical aid, thus they are clearly identifiable. With more states equalizing categorical programs over the past decade, this type of adjustment has become more difficult. More problematic, however, deletion of funding related to vertical equity factors limits the analysis of disparity to a horizontal equity analysis but vertical equity should not be assumed.

Using a second approach to adjust the revenue (expenditure) variable, these considerations are taken into account. Pupil weights are employed to adjust the revenue variable; such weights stipulate the percentage of additional dollars beyond general education aid that are due to justifiable differences in costs for special need students or districts (schools).[3] When weighted pupils are utilized, supplemental revenue drops out of the calculation and, in essence, both a vertical and a horizontal analysis occurs simultaneously. Therefore, this type of analysis is preferred.

Using this approach, "weights" are derived based on the ratio of excess costs to general program costs; they are driven off the cost of the general education program, which is the level of 1.00 in the system; they are empirically derived. Thus, a child with special educational needs would be weighted based on research-based cost differentials that establish the ratio of excess costs to general program costs. For example, a child enrolled in a program that costs half again as much as the general education program is weighted 1.5; another 50 percent in funding is required. The use of weighted pupils in an analysis of equity, however, raises several important questions: What weights should be utilized for differing student configurations and needs, and what factor should target eligible students?

Remedial (Compensatory) Education Weight

According to Levin (1989) the costs of compensatory education vary from .25 but on average it would cost "half again as much" to provide remedial education as it does to provide general education. This would equate to a weight of .5. The estimate is based on the costs of several remedial programs, as reported elsewhere, such as Success for All, Accelerated Schools, and Headstart. A reanalysis of empirical data drawn from a 50-state survey on the costs of programs for special populations (Verstegen & McGuire, 1991) concurs. It indicates that compensatory education programs cost, on average, at least 40 percent beyond general education; this equates to a weight of .4.[4] Thus, remedial costs may vary, but available research indicates that they are, on average, 25-50 percent beyond the cost of the general education program. Students weighted by this factor may be selected by criterion used to distribute aid for the federal remedial education program (ESEA, Title I)—the number of children in poverty by district as measured by the decennial census. An alternative criterion is the number of free lunch recipients, a factor that is also available for schools and states.

Limited English Proficiency Weight/Bilingual Education

Research has not established a single cost estimate for limited English/ bilingual education programs. Like remedial education programs, costs depend on the size of the population, the programs provided, and the goals of those programs. However, Parrish and Matsumoto (1989) suggest an additional amount of 20 percent above the basic grant, based on research estimates (Parrish, 1994). A reanalysis of comprehensive 50-state data (Verstegen & McGuire, 1991) provides a similar but slightly higher estimate of 28 percent. A weighting of .3 is suggested, however, as the data err on the low side.[5]

Special Education Weight

Research has consistently shown that the cost of educating children with disabilities is, on average, double the cost of educating children without disabilities (Rossmiller, Hale, & Frohreich, 1970). Over time, these excess costs of providing special education and related services (costs in "excess" of regular education) have increased slightly from the previous estimate of two times the cost of educating nondisabled children to 2.3 times such costs (Moore, Strang,

Schwartz, & Braddock, 1988; Chaikind, Danielson, & Baven, 1993; Parrish & Verstegen, 1994). When disaggregated, however, it is important to note that special education costs vary widely, both according to student needs and the intensity of programs and services provided to meet these needs. Research shows that costs are relatively higher for low-incidence disabilities such as deaf and blind impairments; and costs are generally lower for high-incidence disabilities, such as learning disabilities and speech (Chaikind, Danielson, & Baven, 1993). Thus, the broader the level of analysis (e.g., federal, state) the more accurate the use of an average weighted estimate.

Multiple weights for special education based on empirically derived estimates can also be calculated and utilized, depending on the quality of state (local) data available. One such approach would provide three tiers of funding adjustments (weights) to reflect the excess costs of providing programs and services to meet exceptional students' needs, based on empirically derived weights for exceptionalities in each category—mild, moderate, and severe (which are utilized as a proxy for the intensity of services required). Weighted averages, based on the number of children in each category are then computed across each of the three funding tiers for all disabilities included in that tier. The three tiers (illustrative) and corresponding weights are:

- *Level I-High Intensity of Services* (Severe Disabilities)[6]—weights students at 2.84 times the cost of general education.
- *Level II-Moderate Intensity of Services* (Mild/Moderate Disabilities)[7]—weights students at 1.77 times the cost of general education.
- *Level III-Special Services* (Mild Disabilities)[8]—weights students at 1.53 times the cost of general education.

Students eligible for special education are those receiving special education and related services as of December 1 of the prior year.

Adjustments for District Costs

According to researchers, school finance equity analysis "should, at the very least, use price-adjusted dollars" (Berne & Stiefel, 1984, p. 280). This allows disparities to be measured in real rather than nominal terms (Chambers, 1979, p. 15). Until recently, however, price adjustments were available in only a handful of states and researchers have been mixed on their uses. However, progress has been made over time and indexes have been developed for all 50 states (cf. McMahon & Chang, 1991; Nelson, 1989; Chambers, 1995), suggesting that, unlike the past, price adjusted dollars might be used more frequently in future fiscal equity studies.

Other legitimate differences in district costs include economies and diseconomies of scale due to differences in district (school/classroom) size; transportation; and different requirements for the "production of safety" (Berne & Stiefel, 1984, p. 15). Although research could inform weightings for size differences and be incorporated into an analysis, this has yet to occur. Thus adjustments for costs are usually made through exclusion (transportation, security) except in the case of price adjustments which could modify the final revenue figure.

A final consideration relates to inflation adjustments. In cross-time studies, measures sensitive to inflation (e.g., range, restricted range) will need to be adjusted to allow real changes in funding to be measured. Various indexes are used depending on the level of the system being examined and the object of analysis (for a discussion, see Verstegen, 1990d) and new indexes have been developed to better target education programs and services (Rothstein & Miles, 1995).

AN ANALYSIS OF EQUAL OPPORTUNITY IN EIGHT STATES AND FOR THE UNITED STATES

In this section cross-time values of equity measures are given for eight states and the United States. Due to space limitations, only major comparative findings are highlighted. Data were obtained from each state department of education over a seven-year period; U.S. data were drawn from documents in the field (NEA, 1995).[9] This information is presented in Tables 6.1 and 6.2. Table 6.1 provides 10 measures of fiscal equity, under the equal treatment of equals principle for children.[10] Table 6.2 presents six measures of the relationship between per pupil revenue and local property wealth, under the equal opportunity principle for children. In Tables 6.3 and 6.4 a comparison of fiscal equity is given for two states, with and without vertical adjustments to the revenue variable. This provides a simultaneous analysis of horizontal and vertical equity and adds the principle of "unequal treatment of unequals" to the analysis. Finally, Table 6.5 provides an analysis of adequacy for one state. Measures used in the analysis are described elsewhere at length (Berne & Stiefel, 1984).

Pupil weighted (state and local) revenue for current operations were the object of analysis, except in Virginia, where adjustments were made to correct for the excess costs of pupils and districts (Verstegen & Salmon, 1989, 1991). Equity within and between states was the focus. All states (for the U.S. analysis) and districts (for each state) were included with the exception of Ohio, where extreme outliers were omitted.

Table 6.1. Cross-State Comparisons of Fiscal Equity for Public School Finance: Univariate Measures

	United States		*Alabama*		*Indiana*		*Missouri*	
Measure:	1980	1990	1980	1993	1993	1994	1980	1990
Range	2,989	5,202	1,023	2,449	2,425	2,468	2,743	7,098
Low	1,229	2,612	924	2,371	3,594	3,306	968	2,654
High	4,218	7,813	1,947	4,820	5,696	5,774	3,712	9,752
Range ratio	3.43	2.99	2.11	2.03	1.58	1.75	3.83	3.67
Restricted Range	1,329	3,124	524	1,035	1,050	1,050	777	3,315
5th%	1,429	2,997	1,133	2,535	3,364	3,431	1,252	3,096
95th%	2,758	6,121	1,657	3,570	4,415	4,486	2,030	6,411
Restricted Range Ratio	1.93	2.04	1.46	1.41	1.31	1.31	1.62	2.07
Federal Range Ratio	0.93	1.04	0.46	0.41	0.31	0.31	0.62	1.07
Coefficient of Variation	0.2062	0.2596	0.1228	0.1249	0.0994	0.1018	0.1737	0.3158
Gini Index	0.0210	0.3208	0.0074	0.0074	0.0048	0.0050	0.0142	0.0432
McLoone Index	0.8448	0.8308	0.9189	0.9287	0.9635	0.9592	0.8769	0.9060
Atkinson Index								
E8	0.8658	0.8229	0.9659	0.9700	0.9739	0.9724	na	na
E10	0.8361	0.7876	0.9528	0.9604	0.9685	0.9666	na	na

(continued)

Table 6.1. (Continued)

	Ohio		Rhode Island		Tennessee		Virginia	
Measure	1981	1900	1990	1992	1960	1989	1988	1990
Range	4,5690	9,166	4,626	6,565	295	2,655	3,844	4,372
Low	1,946	2,480	4,544	4,732	141	1,774	2,654	2,845
High	9,767	11,646	9,170	11,298	436	4,429	6,498	7,220
Range ratio	5.02	4.70	2.02	2.39	3.11	2.50	2.45	2.54
Restricted Range	1,085	2,092	1,446	1,898	160	1,692	2,283	2,665
5th%	1,432	2,736	4,717	4,996	na	na	2,675	3,007
95th%	2,517	4,829	6,163	6,894	na	na	5,177	5,674
Restricted Range Ratio	1.76	1.76	1.31	1.38	2.04	1.93	1.84	1.91
Federal Range Ratio	0.76	0.76	0.31	0.38	0.88	0.82	0.83	0.91
Coefficient of Variation	0.1938	0.2105	0.0950	0.0990	.2301	0.1999	0.2316	0.2734
Gini Index	0.10	0.11	0.05	0.05	0.12	0.11	0.12	0.14
Theil Index	0.0175	0.0203	0.0045	0.0048	0.0252	0.0195	0.0252	0.0348
McLoone Index	0.8819	0.8956	0.9278	0.9261	0.9049	0.8737	0.9099	0.9064
Atkinson Index								
E8	0.9078	0.9074	0.9708	0.9700	0.9089	0.9166	0.8895	0.8653
E10	0.8849	0.8912	0.9630	0.9570	0.8858	0.8917	0.8722	0.8472

Source: Deborah A. Verstegen; University of Virginia, 1996. Includes total state and local revenue unless otherwise noted.

Table 6.2. Cross-State Wealth Neutrality Statistics for Public School Finance: Bivariate Measures

	United States		*Alabama*		*Indiana*		*Missouri*	
Measure:	1980	1990	1980	1990	1993	1994	1980	1990
Correlation	0.50	0.87	0.070	0.514	0.293	0.241	0.71	0.74
Regression	0.25	0.76	0.005	0.2646	0.086	0.581	0.51	0.55
Slope	0.0199	0.0387	0.0002	0.0028	0.0059	0.0053	0.0154	0.0001
Elasticity	0.6136	1.0560	0.0168	0.1340	0.0700	0.060	0.2883	0.1749
F-Ratio	16	156	0.618	45.68	27.59	18.13	562	657
Probability	≤.0002	≤.0001	=.4334	≤.0001	≤.0001	≤.0001	≤.0001	≤.0001
	Ohio		*Rhode Island*		*Tennessee*		*Virginia*	
Measure:	1981	1990	1990	1992	1960	1989	1988	1990
Correlation	0.37	0.52	0.61	0.70	0.65	0.17	0.81	.85
Regression	0.14	0.27	0.37	0.47	0.42	0.28	0.65	0.72
Slope	0.0082	0.0121	0.0029	0.0030	0.0050	0.0045	NA	NA
Elasticity	0.1474	0.1915	0.1205	0.1325	0.1186	0.0470	NA	NA
F-Ratio	95.7	228	19.65	31.35	95.97	4.053	242	330
Probability	≤.0001	≤.0001	≤.0001	≤.0001	≤.0001	≤0.0461	≤.0001	≤0.0001

Note: NA = Not available as wealth measure is an index, rather than property value.

Measuring Equal Treatment of Equals: Study Findings

To quantify the extent to which disparity in per pupil revenue was present cross time in state apportionment schemes, a variety of equity measures were employed and clustered into five conceptual categories of equity: (1) measures of the extremes in the distribution, (2) measures including all observations, (3) measures of the lower portion of the distribution, (4) other measures of equity (Berne & Stiefel, 1984). Using this schema, the index I developed to provide information on observations above the median, would fall into yet another category: (5) measures of the upper portion of the distribution (see Verstegen, 1996).

Measures of Equity Between the Extremes in Revenue Per Pupil

The first conceptual grouping of equity measures includes the range, the restricted range, and the ratios of each. Measures of equity between the extremes in revenue per pupil are sensitive to equal percentage increases and are insensitive to equal additions. Also, not all observations are taken into account—only the highest and lowest. For 1989-1990, the measures of the extremes show that disparities in per pupil revenue averaged $5,200 among the states; they ranged from about $2,400 in Alabama and Indiana to over $9,000 in Ohio. The range ratio indicated that there is about three times more funding available to students in high spending states than to their counterparts in low spending states. Within states, the ratio varied from 1.75 in Indiana to 4.70 in Ohio. Range ratios have decreased over time in five states (Alabama, Missouri, Ohio, Tennessee, Virginia); they have increased in another three states and across all states (Indiana, Rhode Island, Tennessee, the United States).

Excluding five percent of the pupils at each extreme, the *restricted range* ratio (95th and 5th observation) also grew larger over time in all but one state (Alabama); it held steady in another two states (Indiana, Ohio). No state had a restricted range ratio below 1.3; four states had ratios above 1.5 (Missouri, Ohio, Tennessee, Virginia); and ratios above 2.0 were found for the United States overall, and for Missouri.

Measures Including All Observations

The second conceptual grouping of equity measures include the coefficient of variation, the Gini index, and the Theil index. These measures are insensitive to equal percentage increases (inflation) and include all observations.

In 1989-1990 the *coefficient of variation* (COV) was 26 percent for the United States. This means that about two-thirds of all pupils were within 26 percent of the average per pupil revenue amount, and about nine-tenths were within 52 percent of the average. The COV was at or below 10 percent in two states (Rhode Island, Indiana) but increased over time—it showed less equity—in all states except one (Tennessee) where an extended time horizon may indicate the statistic is an artifact of measurement or reflects improved databases. The Gini index also indicated less equity over time both within states and between them, except in Tennessee, Rhode Island, and Alabama (where it held steady). In 1990 it ranged from 0.05 in Rhode Island to 0.15 in Missouri; it was .14 among the states (United States). As the Gini index approaches zero, more equity is shown—this is also true for the Theil index. The Theil index, like the Gini, also increased over time, that is, showed less equity, for all states except Tennessee (see above) and Alabama.

Measures of the Lower Half of the Distribution

The third conceptual grouping of equity measures includes the following three statistics: the McLoone index and the Atkinson index with high values of the parameter E, where $E = 8$ and $E = 10$. These statistics measure equity for the lower portion of the revenue distribution only. The McLoone index showed less equity over time, both between and within most states with the exception of Ohio, Alabama, and Missouri where slight improvements were found. In 1990 the McLoone index was 0.83 among all states; within states it ranged from a low of .87 in Tennessee to a high of .96 in Indiana (1993). As the McLoone increases, equity increases. This is also the case for the Atkinson index. The

Atkinson index with high values of the parameter E showed less interstate and intrastate for most states—except Tennessee and Alabama, when $E = 8$. With $E = 10$, equity improvements were noted in Tennessee and Ohio but an erosion of equity was evidenced for other states examined and for the United States overall.

Other Equity Statistics

The final conceptual category of equity includes a single measure, the Federal range ratio. In 1989-1990 the Federal Range Ratio was 1.04 among all states. It ranged from .31 in Rhode Island and Indiana to 1.07 in Missouri, thus mirroring results found for the restricted range ratio (minus 1.0). Because these ratios are in excess of the permissible variance allowed by the U.S. Department of Education under 34 CFR 1, Sec. 5 (d)(2), if the state legislature in these states desired to take into consideration the federal impact aid funds prior to allocation of state aid, the state would be prohibited from doing so.

Measures of the Upper Half of the Distribution

The fifth conceptual grouping of equity measures include the index that I developed that measures equity for the upper portion of the revenue distribution only. It is based on the ratio of funding available for pupils above the median, to funding those pupils would receive if they were at the median. The greater the amount of revenue that is allocated to students on whom spending is greater than the median funding level, the larger the index. It is discussed at length elsewhere (Verstegen, 1996). The index for the upper portion of the revenue distribution showed that greater dispersions in the top half of the distribution occurred over time in Alabama, Missouri and Ohio, and Virginia, but held steady in Indiana.

Bivariate Statistics and the Measurement of Wealth Neutrality: Study Findings

Wealth neutrality statistics show the relationship between a locality's per pupil revenue for education and its *ability to pay* for education through local sources, that is, fiscal capacity or wealth. Theoretically, there should be no relationship between these two factors. Statistics measuring these relationships include correlations regressions, slopes, and elasticities.

Pearson Product-Moment *correlations* showed moderately strong to high relationships between wealth and revenue in 1990 both between states (United States) and within states—with the exception of Tennessee ($r = .17$, 1990) and Indiana ($r = .29$, 1993). Cross-time changes indicated these relationships strengthened over time. Between states the correlation between wealth and revenue was high in 1980 ($r = .50$), but grew in 1990 by over one-half ($r = .87$, 1990). Within states, strong associations were also evidenced (Alabama, $r = .51$; Missouri, $r = .74$; Ohio, $r = .52$; Rhode Island, $r = .70$; Virginia, $r = .85$). These data indicate an increasingly nonwealth-neutral distribution and an erosion of equity in general between and within the states examined in this study. This was underscored by the regression coefficients. The *regressions* showed that in 1990 between one-half to two-thirds of a locality's revenue for education could be accounted for by its respective ability to pay (wealth) for all but two states (Alabama and Tennessee). The coefficients were significant and robust in all states (except Alabama where low levels of revenue overall were found). The magnitude of the relationship between wealth that is, the ability to pay, and revenue, as indicated by the *slopes*, tended to be below 0.01 within states. Thus, a $100 change in wealth was associated with a $1 change in revenue. The *elasticity*, like the slope, indicated that in 1990 the magnitude of the relationship between wealth and revenue varied; it ranged from .06 in Indiana to 1.07 across all states (United State).

Measuring Equal Treatment of Equals and Unequal Treatment of Unequals: Study Findings

To quantify the extent to which disparity in per pupil revenue varied cross time, when adjustments were made to the revenue variable to correct for vertical factors—special needs students—data were analyzed with and without adjustments which were made by deletion (Tables 6.4 and 6.5). Interestingly, Table 6.4 shows that univariate measures for these states showed less variance overall when adjustments for vertical equity factors were taken into account, but the changes were not large. For example, in Rhode Island, the range in revenue fell from $6,565 per pupil to $6,517 per pupil; the COV dropped from .099 to .083, and the McLoone increased from .926 to .951. Thus, the statistics showed that revenue inequalities were tempered somewhat as a result of improvements in measurement. However, as shown in Table 6.5, the relationships between wealth and revenue strengthened when adjustments were made to the revenue variable, as depicted by the bivariate measures. Thus, equal opportunity showed further erosions. For Rhode Island, for example, the association between wealth and revenue as captured by the regression coefficient increased from .479 to .507; thus wealth could explain over half the variation in revenue among school districts in the state after adjustments, previously it could explain 48 percent of the variation in revenue. Thus, adjustments may restrict the dispersion in univariate measures as evidenced by multiple measures, but allow the effects of illegitimate associations to be enhanced, thus indicating further movement away from equal educational opportunity.

Table 6.3. Weighted and Unweighted Fiscal Equity Statistics for School Finance

	Indiana 1993		Rhode Island 1992	
Measure	*Revenue*	*Revenue Adjusted*	*Revenue*	*Revenue Adjusted*
Range	$ 2,525	$ 2,208	$ 6,565	$ 6,517
Lowest (0%)	$ 3,594	$ 3,066	$ 4,732	$ 4,572
Highest (100%)	$ 5,696	$ 5,274	$ 11,898	$ 11,088
Restricted Range	$ 1,050	$ 851	$ 1,898	$ 1,414
5th%	$ 3,364	$ 3,117	$ 4,996	$ 4,842
95th%	$ 6,894	$ 6,256	$ 6,894	$ 6,256
Federal Range Ratio	0.3122	0.2730	0.3799	0.2920
Coefficient of Variation	0.0994	0.0910	0.0990	0.0831
Mean	$ 3,751	$ 3,435	$ 5,956	$ 5,568
Geni Index	0.0533	0.0485	0.0530	0.0439
Theil's Index	0.0048	0.0040	0.0048	0.0034
McLoone Index	0.9635	0.9574	0.9261	0.9514
Median	$ 3,595	$ 3,336	$ 5,972	$ 5,496
Atkinson's Index				
I8	0.9739	0.9574	0.9666	0.9774
I10	0.9685	0.9732	0.9570	0.9709

Table 6.4. Weighted and Unweighted Wealth Neutrality Statistics for School Finance, 1993, 1994

	Indiana (1993)		Rhode Island (1992)	
Wealth-Assessed Valuation	*Revenue*	*Revenue/ Adjusted*	*Revenue*	*Revenue/ Adjusted*
Correlation (*r*)	0.6926	0.7122	0.2929	0.3301
Regression (r^2)	0.0858	0.1059	0.4797	0.5072
Slope	0.0059	0.0061	0.0030	0.0024
Elasticity	0.0700	0.0715	0.1325	0.1144
F-Ratio	27.59	35.95	31.35	34.99
Probability	≦0.0001	≦0.0001	≦0.0001	≦0.0001

Table 6.5. Adequacy Issues: A Comparison of Alabama to Other States on Multiple Indicators

	Total:	*Rank:*
Variable:	*States Reporting*	*Alabama*
• Gross State Product Per Capita, 1980, (Constant)[1]/	50	43
• Gross State Product Per Capita, 1989, (Constant)[1]/	50	43
• Gross State Product Per Pupil, 1990, (Current)[1]1/	50	41
• Gross State Product Per Pupil, 1989, (Current)[1]/	50	41
• Education Aid (Federal, State, and Local), 1980[2]/	50	49
• Education Aid (Federal, State, and Local), 1990[2]/	50	50
• Education Aid, 1990, (State & Local)[2]/	50	49
• Total Mathematics Test Scores, National Assessment of Ed. Progress (NAEP), Percent of 8th Graders Scoring at least at 200[3]/	35	33
• Mathematics 300—percent 8th graders scoring at least 300 in math (NAEP)[3]/	35	34 (4-way tie)
• Mathematics 250—percent 8th graders scoring at least 250 (NAEP)[3]/	35	33
• Elementary Teacher's Salary, 1990[4]/	50	38
• Secondary Teacher's Salary, 1990[4]/	50	40
• Percent Adults Aged 25+ With 1-3 Years High School (1980)[5]/	50	42
• Percent Adults Aged 25+ With 1-3 Years College[5]/	50	44
• Percent Adults Aged 25+ With 4 or More Years College[5]/	50	47
• Percent Mothers Less Than 19 Years With Less Than 12 Years of Schooling, 1986[7]/	50	3 (2-way tie)

Note: 1 Bureau of Economic Analysis (1992). Gross State Product in Current & Constant Dollars. Washington, DC: U.S. Department of Commerce (1982 = 100). 2 Data year, 1990. National Education Association, (1990). Digest of Education Statistics. West Haven, Author. 3 U.S. Department of Education (1991). Digest of Education Statistics, 1991. Washington, DC: U.S. Government Printing Office. Table 116, p. 118. Note, level 200 generally tests reasoning and problem solving skills, level 250 tests two step problem solving, level 300 tests fractions, decimals, percentages, geometry, and algebra, level 350 tests geometry, algebra, statistics, and probability. 4 Data year 1990. Computed from aggregate data: U.S. Department of Education (1991). Digest of Education Statistics, 1991. Washington, DC: U.S. Government Printing Office. 5 Data year, 1980. U.S. Department of Education (1991). Digest of Education Statistics, 1991. Washington, DC: U.S. Government Printing Office.

Measuring Adequacy—One Approach: Study Findings

Because measures of equity are insensitive to the level of funding available, analysis was undertaken to compare the relative position of one state, Alabama, to all states, on various input and output measures (for a more thorough research design, see Verstegen & Whitney, 1995; Verstegen, 1996). Recall that Alabama had relatively low variances as depicted by the univariate measures. It was postulated that the distribution overall was inadequate, due to low levels of funds. The data showed that Alabama ranked at the bottom on all measures. It was 50th out of 50 states in the amount of funding devoted to public elementary and secondary education (1990); it was 33rd out of 35 states and

34th out of 35 states on two measures of mathematics proficiency on the National Assessment of Education Progress; and it was 47th out of 50 states on the percent of adults aged 25 years and older that completed college. Alabama's ability to fund education was also low, as measured by the gross state product (GSP); it ranked 43rd out of 50 states on GSP per capita and 41st on GSP per pupil. But on total it ranked education aid 50th out of 50 states; and teacher's salaries were in the bottom 10 percent of all states. Thus, compared to all states, Alabama was in the bottom 10-20 percent on these input and output measures.

SUMMARY

This chapter has discussed conceptual and measurement issues related to the analysis of equal educational opportunity; and provided equity analyses for eight states and the United States. The study included 15 statistical measures of equity; analysis was undertaken with and without adjustments for legitimate variations in costs for comparison purposes, and an ancillary, comparative adequacy analysis was provided for one state. In addition, data for empirical derivation of cost differentials were presented and associated weights produced.

As discussed in the chapter, advances in databases that make school-level data more available, in methodology that permit simultaneous vertical and horizontal equity analysis through the use of empirically derived cost differentials, and new measures of equity—when taken into account by researchers—should enhance future assessments of equal opportunity and prove a better guide to public policy aimed at fairness and justice issues for children and youth in America's public elementary and secondary schools. In addition, improvements in equity research should inform efficiency analysis, as the "master input" variable, revenues (expenditures), is common to both analyses. The assumption is that equity and efficiency analysis will work better together than either would alone in addressing issues related to increasing productivity for all children in America's schools.

NOTES

1. In school-level analysis no relationship should exist between revenue and family income, for example, poverty; in interstate analysis gross state product may be used for the wealth variable.

2. For a discussion of difficulties in productivity analysis when total revenues are used (rather than adjusted revenue) see, Baker,

3. Weights can also be used for size corrections (see Verstegen, 1990c); however transportation and security are best deleted.

4. There was a 100% response rate. Data are federal and state funds for remedial education programs over federal, state, and local funds—so err on the low side because local funding for remedial programs was not included. Average funding from all sources in 1987-88 was $3,962 (NEA, *Estimates of School Statistics*, 1995). See Verstegen & McGuire (1991) Table 3 (funding)

and Table 4 (population). Thus, without including locally expended funds for remedial education, the weight is .375. A weight of .4 is suggested to account for local funds, but actual costs may be higher, and vary from site to site.

5. There was a 100% response rate. Reanalysis of data includes federal and state funds over federal, state, and local funds, so the weight errs on the low side because local funding for LEP programs was not reported but is included in the divisor. See Verstegen & McGuire (1991) Table 3 (funding and Table 4 (population reported). Average funding from all sources in 1987-88 was $3,962 (NEA, *Estimates of School Statistics*, 1995). Thus, without including locally expended funds for LEP, the weight is .278. A weight of .3 is suggested to account for local funds, but actual costs may be higher, and vary from site to site.

6. Mentally impaired (severe); emotionally/behaviorally impaired; deaf/blind; hearing impairments/deaf; visual impairments/blind; traumatic brain injury; autism.

7. Mentally impaired (mild/moderate); other health impaired; orthopedically impaired.

8. Speech; specific learning disabled.

9. Data included state and local revenue; they were obtained under subcontract from the department of education in each state.

10. Taxpayer equity provides an alternative method of investigation (see Berne & Stiefel, 1984). However, in an analysis of opportunity, children are the critical focus.

REFERENCES

Baker, K. (1991). Yes, throw money at schools. *Phi Delta Kappan, 72*, 628-631.

Berne, R., & Stiefel, L. (1984). *The conceptualization and measurement of equity in school finance.* Baltimore, MD: Johns Hopkins University Press.

________ . (1993). Measuring equity and efficiency at the school level: The finance perspective. Paper presented at the American Education Policy and Management annual conference, October.

Berne, R. (1994). Preface. In R. Berne & L.O. Picus (Eds.), *Outcome equity in education.* CA: Corwin Press.

Chambers, J. (1995, October). *Public school teacher cost differences across the United States.* (NCES 95-758). Washington, DC: U.S. Department of Education.

Chaikind, S., Danielson, L.C., & Baven, M.L. (1993). What do we know about the costs of special education? A selected review. *Journal of Special Education, 26* (4), 344-370;

Clune, W.H. (1992). New answers to hard questions posed by *Rodriguez*: Ending the separation of school finance and education policy by bridging the gap between wrong and remedy. *Connecticut Law Review, 24* (3), 721-753.

Cooper, B.S. (1994). Making money matter in education: A micro-financial model for determining school-level allocations, efficiency and productivity. *The Journal of Education Finance, 20* (1), 66-87.

Fortune, J., & O'Neil, J. (1994). Production function analysis and the study of education funding equity: A methodological critique. *The Journal of Education Finance, 20* (1), 21-46.

Geske, T.G., & LaCost, B. (1990). The student equity effects of the public school finance system in Louisiana. *Economics of Education Review, 9* (1), 55-65.

Hartman, W.T., & Haber, T.R. (1981). *School finance reform and special education.* Eugene, OR: ERIC No. ED 207 165.

Hertert, L., Busch, C., & Odden, A. (1994). School financing inequities among the states: The problem from a national perspective. *Journal of Education Finance, 19* (3), 231-256.

Jefferson, T. (1813) (1959). A letter from Thomas Jefferson to John Adams, October 28. In L.J. Cappon (Ed.), *The complete Adams-Jefferson letters* (pp. 387-392). Chapel Hill: University of North Carolina Press.

Levin, H.M. (1989). Financing the education of at-risk students. *Educational Evaluation and Policy Analysis, 11* (1) 47-60.

McMahon, W.W., & Chang, S. (1991, April). Geographical cost of living differences: Interstate and intrastate, update, 1991. MacArthur/Spencer Series (No. 20). Normal, IL: Center for the Study of Educational Finance, Illinois State University.

Moore, M.T., Strang, E.W., Schwartz, M., & Braddock, M. (1988). Patterns in special education service delivery and cost. Contract No. 3000-84-0257. Washington, DC: Decision Resources Corporation.

National Education Association. (1995). *Estimates of school statistics.* New Haven, CT: NEA Library.

Odden, A., Berne, R., & Stiefel, L. (1979). *Equity in school finance.* Report No. F79-9. Denver, CO: Education Commission of the States.

Parrish, T. (1994). A cost analysis of alternative models for limited English students in California. *Journal of Education Finance, 19* (3), 256-278.

Parrish, T., & Matsumoto, C.S. (1989, February). *Disparities in public school district spending, 1989-90.* NCES 95-300. Washington, DC: National Center for Education Statistics, U.S. Department of Education.

Parrish, T.B., & Verstegen, D.A. (1994, June). *Fiscal provisions of the individuals with disabilities education act: Policy issues and alternatives.* Policy Paper Number 3. Palo Alto, CA: Center for Special Education Finance, American Institutes for Research.

Rossmiller, R.A., Hale, J.A., & Frohreich, L.E. (1970). *Educational programs for exceptional children: Resource configurations and costs.* Special Study No. 2. Madison, WI: National Education Finance Project;

Rothstein, R., & Miles, K.H. (1995). *Where's the money gone? Changes in the level and composition of education spending.* Washington, DC: Economic Policy Institute.

Speakman, S.T., Cooper, B.S., Smapieri, B. May, J. Holsomback, H., & Glass, B. (1996). Bringing money to the classroom: A systemic resource allocations model applied to the New York City schools. In L.O Picus & J.L. Wattenbarger (Eds.), *Where does the money go? Resource allocation in elementary and secondary schools.* CA: Corwin Press.

Summers, A.A., & Wolfe, B.L. (1967). Do schools make a difference? *The American Economic Review, 67*, 639-652.

Thurow, L.C. (1985). *The zero-sum solution.* New York: Simon and Schuster.

Verstegen, D.A. (1987). Equity in state education finance: A response to *Rodriguez. Journal of Education Finance, 12* (3), 315-330.

________ . (1990a). Introduction. In D.A. Verstegen & J.G. Ward (Eds.), *Spheres of justice in education* (pp. i-vii). New York: Harper Business.

________ . (1990b). Invidiousness and inviolability in public education finance. *Educational Administration Quarterly, 26* (3), 205-234.

________ . (1990c). Efficiency and economies of scale revisited: Implications for financing rural school districts. *Journal of Education Finance, 16* (2), 159-179.

________ . (1990d). Fiscal policy for education in the Reagan administration. *Education Evaluation and Policy Analysis, 12* (4), 355-373.

________ . (1992). An interstate equity analysis of funding for education: Problems and prospects. Paper presented at the Annual Conference of the American Policy and Management Association, Denver, CO, October.

________ . (1994). Equity and efficiency in the provision and reform of American schooling. *The Journal of Education Finance, 20* (1), 107-131.

________ . (1996). Reforming American education policy for the twenty-first century. In J. Cibulka & W.J. Kritek (Eds.), *Coordination among school, families and communities: Prospects for educational reform.* New York: SUNY Press.

———. (1996). Concepts and measures of fiscal inequality: A new approach & effects for five states. *Journal of Education Finance, 22* (2), 145-160.

Verstegen, D.A., & McGuire, C.K. (1991). The dialectic of reform. *Educational Policy, 5* (4), 386-411.

Verstegen, D.A., & Salmon, R.G. (1989). The conceptualization and measurement of equity in school finance in Virginia. *Journal of Education Finance, 15* (2), 205-228.

———. (1991). Assessing equity in Virginia school finance: Cross-time comparisons. *Journal of Education Finance, 16* (4), 417-431.

Verstegen, D.A., & Whitney, T. (1995, November). *Financing public education: Changes from 1965 to 1995–The role of the courts.* Washington, DC: American Public Policy and Management Conference (APPAM).

Chapter 7

TRACING SCHOOL-SITE EXPENDITURES:

EQUITY, POLICY, AND LEGAL IMPLICATIONS

Sheree T. Speakman, David C. Bloomfield,
Bruce S. Cooper, Jay F. May, Robert A. Sampieri,
Hunt C. Holsomback, Larry Maloney, Judith Nappi, and
Deborah Rosenfield

Since *Serrano* v. *Priest*[1] in 1971 in California, the search for greater equity in education resources has been a prominent feature of the American political and legal landscape. Over the last quarter century, an additional 39 states (following California in 1971) have fought similar court battles over school finance equity. Of the 39 states, 26 witnessed legal rulings recognizing the inequalities of school district resources—usually requiring legislatures to increase state aid to the poorer districts and even to "cap" the spending of the property-rich districts (see Burrup, Brimley, & Garfield, 1996, pp. 231-233;

Advances in Educational Productivity, Volume 6, pages 149-193.

ISBN: 0-7623-0082-5

Wise, 1967; Coons, Clune, & Sugarman, 1989). As the cases widened to include other equity issues, the rulings served as a backdrop to the work supporting the Education of All Handicapped Children Act of 1973, establishing standards for equity in educational opportunities benefiting physically, emotionally, and intellectually challenged children.

Extensive litigation has helped alleviate funding disparities between certain school districts and among students with different resource needs. However, many children still attend schools in districts that receive funding considerably below acceptable state and national norms. As we continue to explore this reality, an analytical shift is occurring, from the examination of funding disparities between *school districts* to the examination of disparities between *individual schools*. This development is consistent with recent trends in school governance and management, where school leaders increasingly exercise significant control over management and to a lesser degree, financial decision making at the school site. Indeed, this "devolution revolution" is consistent with the new reality of education. Everyone agrees that real change in teaching and learning must be implemented, not in school district or state education department offices, but in schools themselves.

Just as we are refocusing our efforts for reform, moving from school districts to schools, so too must we adapt and sharpen our analytical methods to evaluate progress. In the field of education equity research, perhaps the most important theoretical work on the measurement of interdistrict equity in school finance was pioneered by Robert Berne and Leanna Stiefel (1984). These authors distinguish spending equity among *equal* pupils (horizontal equity) from expenditures among students with different problems and needs, the "unequal treatment of unequals" (vertical equity).

Importantly, Berne and Stiefel provided numerous statistical concepts and methods for measuring the distribution of resources to school districts. Their work has made it possible to measure progress (or the absence thereof) toward greater equity—measurements used as evidence in court litigation and in the monitoring of the aftermath of court decisions and subsequent legislation. Since its introduction, Berne and Stiefel's collection of equity measures has become the cornerstone of school finance research on equity. However, these measurement standards were developed with the school district as the unit of analysis, with the baseline assumption being that resources within a district are distributed equitably. New methods, tools, and technologies enable us to test this assumption and to examine whether a new approach to equity, focusing on the school site rather than the district, is more robust and will yield different results.

In this chapter we focus first on the limitations of district-level expenditure analysis. We then document the intersection of traditional equity research and new developments in school-site expenditure analysis. We employ not only a shift from district-wide to school-site analysis of school finance equity, but also a much more detailed presentation of resource expenditure information by location, function, and educational program. In addition, the spending differentials by school levels (elementary, middle, and high schools) are calculated. Finally, with this plethora of new school-level information in hand, we consider the policy and legal ramifications of school-site data analysis.

LIMITATIONS OF DISTRICT-LEVEL ANALYSIS

During the 1990s education policies have increasingly emphasized school-site budgeting, management, decision making, and accountability, and are seen by many as a strong antidote to the uniformity, rigidity, and apparent failure of many district-oriented solutions. As individual schools become the center of attention, states and school districts often intervene directly in ailing schools whose students consistently score below par on standardized performance indicators.

As an example of the reform of school governance and control, the Illinois General Assembly created school-site, elected Local School Councils (LSCs) in the Chicago Public Schools in 1988. LSCs are empowered to hire and fire their school principals, oversee school policies, and adopt a budget for their discretionary dollars. While elected LSCs have experienced varying degrees of success, clearly their members have the influence and authority to create policy and initiate change; they are in a position to support meaningful improvement rather than to tolerate an ineffective status quo.

Educators themselves now understand that unless key decisions are made at the school, departmental, and classroom levels, programs and curricula will not change fast enough to maintain alignment with changing classroom environments and pupil populations. Furthermore, policymakers are demanding greater scrutiny of school productivity in support of higher achievement scores. The definition and measurement of productivity in actual practice should now become more commonplace with improved site-level data on spending, programs, classroom activities, and academic results.

With the proliferation of school-based policy and practice, prominent school finance scholars recognize that district-level equity analysis no longer provides enough detail of the variance in school finance to be meaningful. Allan R. Odden (1992), for example, writes that "moving education finance from District to School finance focuses education funding on the organizational unit—schools—responsible for producing bold new student performance levels embodied in the country's education goals" (p. 336).

Transition to School-Site Analysis

Efforts to make the transition to school-site analysis have had varying degrees of success for several reasons. First, a number of specific expenditures for pensions and benefits, capital formation, debt service, and transportation may not be included in traditional, district-prepared data sets because these expenditures are sometimes paid by other city, county, and state agencies. Therefore, a traditional analysis of the 1,286 public schools in the 24 county-defined districts in Maryland, for example, would tend to understate total 1994-1995 education expenditures by $395,405,701, because the amount of retirement and pension system expenditures paid by Maryland on behalf of the districts is not reflected in district general ledgers (for a state with 790,938 students in 1994-1995, this equates to $499.92 per pupil). A traditional analysis of the 91 school districts in South Carolina would tend to understate total school year 1993-1994 expenditures by $51,013,952, because specific expenditures for testing, textbooks, and transportation paid by the State Department of Education are not reflected in district general ledgers (in a state with 609,008 students in 1993-1994, this equals $83.77 per pupil).

Current district expenditure reports are the product of evolution—they have been modified to answer new questions, one question at a time. As a result, the framework for analyzing district expenditure data has not necessarily evolved to include 100 percent of the dollars in standard issue report—making it virtually impossible to demonstrate that the sum of the parts equals the whole. For its part, school-site analysis must include 100 percent of the dollars that are invested at the school level, and be disciplined, so that the sum of the its parts always equals the whole. In this way, the answers to ad hoc, "puzzle piece" questions, when connected with answers to all of the puzzle pieces, will produce a complete, and therefore comparable, picture of district and school expenditures.

Second, district data sets and standard reports are designed around regulatory accounting, not managerial accounting. While the data produced by districts is always voluminous, its regulatory compliance orientation is designed to respond to specific questions related to the fiduciary responsibility of a state or federal agency, rather than to answer a district's or researcher's questions about managing the administrative and pedagogical aspects of a district or school. As regulatory requirements have evolved, districts have modified their general ledger systems and student systems to respond to the specific questions asked. Few districts have had the mandate, time, and money to step back and rethink their future information needs, or to implement the latest relational database technologies in a system that has adequate flexibility to serve both the needs of compliance reporting for the district and effective management practice in schools. Therefore, while the detail needed to answer many of the financial questions being asked today may be found in a district's general ledger, each question asked often requires additional programming to extract the information and present it in a report. Since making management decisions is a multidimensional process which considers numerous questions simultaneously and iteratively, district data systems fall seriously short of providing "actionable information" about school-site activities.

Third, financial compliance information aggregates expenditures by fund and by point of management control, rather than at the location where resources are consumed—resulting in expenditures that actually benefit schools (i.e., food service, transportation, and staff development) being reported as if they benefited central administration. School-site analysis gains from two specific accounting tools: an accurate and efficient general ledger system and a management information system. With both systems the general ledger should then be used to record expenditures by fund and by point of fiduciary control, while the management information system should input expenditures according to usage and benefit (i.e., where consumed).

In the private sector these two tools are sometimes packaged as separate systems, a financial accounting system and a cost accounting system, each underpinned by process methodology and accounting practice. In public education the majority of districts utilize only a financial accounting system. Without an efficient and systematic means to track and allocate expenditures benefiting school sites, districts are limited in their ability to provide site-based information for school-based management.

Fourth and last, most states and districts do not have a consistent and comparable methodology for grouping and categorizing expenditures for school sites. A variety of data sources are used to answer commonly received inquiries on schools—as a result, the answers differ from state to state and district to district to questions such as: What should be charged to a school versus a central office? Should there be a distinction between operating costs and nonoperating costs? What should be included in the staff development category? What are the elements of the largest category of spending known as function 1000, instruction? Should expenditures be arrayed by restricted fund category (e.g., bilingual restricted funds), by total program spending (e.g., general funds and restricted funds expended on bilingual programs), or both? Although individual ledger transaction codes are often similar from state to state and district to district, the rules for summarizing this information differs. There are even differences in what is considered the total dollar amount in a district.

Despite these shortcomings, a number of researchers are attempting to collect and analyze school-site information. For example, in a recent study (reviewed in draft form and not yet released) conducted by the Pelavin Research Institute, Joel Sherman (1996) examined school-site spending in Ohio and Texas. The study explored the distribution of resources between central offices and school sites, as well as the per pupil expenditures for direct instruction and other functions across different schools and grade-level groupings.

Pelavin faced significant field-work limitations. For example, because the expenditure from the sample states of Ohio and Texas were not fully attributed to the school level, certain of the report's conclusions were misleading. An example would be that without all student expenditures attributed to schools, the percentages calculated for classroom instruction were effectively overstated. Additionally, because Pelavin did not present aggregate district-level expenditure amounts in the analysis, a reader is unable to determine the percent of each district's spending that was allocated to the schools evaluated in the study.

Despite these data-set limitations, however, the Pelavin study identifies interesting patterns of expenditures and warrants continued analysis. The study found that in both states, elementary schools dedicated a higher percentage of resources to classroom instruction (61.6% in Ohio; 77.8% in Texas) than did high schools (59.4% in Ohio versus 76.4% in Texas). The preponderance of other school finance research, focusing on total district expenditures rather than on school expenditures, has shown that high schools spend more dollars in total than do elementary schools. Combining the two realities, the usage analysis illustrates in empirical terms that while high schools expend more resources overall (as a percent of district per pupil expenditures) than do elementary schools, high schools also spend a lower percentage of resources in the classroom.

These patterns are understandable, that is, high schools tend to have larger school plants, higher management charges for more complex departmentalized academic schedules, and more extensive extracurricular activities, among other differences. However, the benefit to learning and achievement of driving more dollars into the high school classroom, and in fact all classrooms, needs a fresh evaluation. To investigate resource allocation more specifically at the school level, school-level data for a significant number of districts, intra- and interstate will be needed.

As mentioned earlier, the Pelavin study treated all schools in Ohio and Texas as units of analysis but improperly ignored the costs associated with managing the overall operation of school districts. These costs vary widely from district to district and thus cannot be ignored. Putting $3,000 into the classroom for instruction may not affect educational delivery the same way in different districts. As an example, a district that expends $6,000 per pupil and is putting $3,000 into the classroom, is allocating 50 percent to the classroom. Another district that spends $3,000 per pupil for instruction but $8,000 per pupil overall, is effectively spending only 37.4 percent in the classroom. An appropriate question is: once the dollars and percentages are calculated for teaching and learning, can we determine the impact (if any) of resources on student's achievement?

Berne's (1994) work in the New York City Public Schools was another study of school-site spending, analyzing instructional costs in each of the system's 32 community school districts. This study's major limitation was the inability to discriminate between expenditures for instruction that would reach the schools themselves as compared to the costs retained by the district office to manage instructional programs. It was not a focus of the study to trace resources to each of the 1,000-plus elementary and middle-level schools in those 32 community school districts. However, NYC Public Schools have just announced their intent to implement site-based budgeting and reporting across the system within a three- to five-year period. This policy promises the opportunity to disaggregate administrative and operational expenditures for all schools and district offices.

Linda Hertert (1995) voices a concern about the long-time usage of expenditure analysis focused on districts instead of schools: she explains that "the use of district-level data, however, may not be appropriate if the goal is to measure, and thereby to promote, a more equitable allocation of funds among pupils. That is, it may be incorrect to assume that the distributional and/or spending patterns across districts mirror those among the schools that children attend" (p. 71). While Hertert based her research on the comparative analysis of spending by district and within districts in California, her point may also hold for a nation of schools.

EQUITY RESEARCH AND SCHOOL-SITE ANALYSIS

This research addresses the question of whether educational resources are indeed distributed equitably across schools within a district, by function and by program. Toward this end, Coopers & Lybrand L.L.P. and the U.S. Chamber of Commerce's Center for Workforce Preparation created a model to analyze three dimensions of spending: function, educational program, and type of school. Pioneered by Dr. Bruce S. Cooper and his colleagues at Fordham University's Graduate School of Education, this methodology was later expanded into a personal computer software package called In$ite™—The Finance Analysis Model for Education™ (In$ite), which arrays and analyzes a school district's expenditures using three-dimensional analysis. We demonstrate the application of this methodology to equity concerns by means of two district case studies.[2]

Methodology—Preparing School-Level Data

In$ite is a management information system that mitigates the four challenges of transitioning to school-site analysis noted earlier. In$ite imposes a strict financial language and methodology for evaluating district- and school-level data, which ensures a valid three-dimensional analysis, and reconciles to a school district's audited financial statements. Despite the trend toward school-based reform initiatives, school districts, not schools, often have taxing authority, receive state and federal funds, and are charged with the responsibility to report budgets and audited financial statements to the state.

Reconciling In$ite's information to audited financial information is an important step in the model's methodology to assure data integrity and credibility.

In$ite requires that 100 percent of a school district's total expenditures be analyzed, including both expenditures directly attributable to individual schools, and those accounted for centrally (some of which may then be allocated). In$ite also incorporates expenditures made on behalf of the district by other entities (e.g., city, county, states, and federal agencies). Rather than rely on state-specified regulatory reports, which contain only highly aggregated school district financial information, In$ite imports a school district's detail general ledger expenditure information into a relational database, from which "actionable" management information can be produced. All subsequent aggregation and analysis are performed using a district's original "electronic" data as the source.

The three-dimensional framework of the relational database automatically assures that the sum of the parts equals the whole, regardless of what subset of data is analyzed. Each general ledger line item, for each location, is coded to capture the dimensions of the model: by function, by educational program, and by school level or central office site (Elementary Schools, Middle Schools, High Schools, Alternative Schools, Other Schools and Non-Schools). The extensive standard reports (in excess of 280 reports, plus additional reports for each school) are tailored to respond to questions from specific stakeholders: some answer parent questions, administrator questions, legislator questions, researcher questions, taxpayer questions, principal questions, and board member questions.

By Function

First, an In$ite user assigns each district expenditure to one of 32 separate detail functions; the software then aggregates these detail functions into 15 sub-functions and five summary functions. The 32 detail functions are numbered sequentially below by sub-function and summary function:

1. **Instruction:** Includes human and physical resources expended to teach students, face to face in the classroom.
   (1) Face-to-Face Teaching
      1. *Instructional Teachers*
      2. *Substitute Teachers*
      3. *Instructional Paraprofessionals*

   (2) Classroom Materials
      1. *Pupil-Use Technology and Software*
      2. *Instructional Materials, Trips, and Supplies*

2. **Instructional Support:** Includes expenditures that support the needs of students, teachers, and programs to prepare them for teaching and learning, generally outside the classroom.
   (1) Pupil Support:
      1. *Guidance and Counseling*
      2. *Library and Media*
      3. *Extracurricular*
      4. *Student Health and Services*
   (2) Teacher Support
      1. *Curriculum Development*
      2. *In-Service, Staff Development, and Support*
      3. *Sabbaticals*
   (3) Program Support
      1. *Program Development*
      2. *Therapists, Psychologists, Evaluators, Personal Attendants, and Social Workers*

3. **Operations:** Includes noninstructional expenditures necessary to provide the infrastructure of the teaching and learning enterprise.
   (1) Noninstructional Pupil Services
      1. *Transportation*
      2. *Food Service*
      3. *Safety*
   (2) Facilities
      1. *Building Upkeep, Utilities, and Maintenance*
   (3) Business Services
      1. *Data Processing*
      2. *Business Operations*

4. **Other Commitments:** Includes expenditures not related to day-to-day instruction, support, or operation of the district in the current year, for this year's student body of the district.
   (1) Contingencies
      1. *Budgeted Contingencies*
   (2) Capital
      1. *Debt Service*
      2. *Capital Projects*
   (3) Out-of-District Obligations
      1. *Parochial, Private, Charter, and Public School Pass-Throughs*
      2. *Retirement Benefits and Payments*
      3. *Enterprise and Community Service Operations*
   (4) Legal Obligations
      1. *Settlements*

5. **Leadership:** Includes expenditures for central office and school-based instructional and noninstructional management professionals.
   (1) School Management
      1. *Principals and Assistant Principals*
      2. *School Office*
   (2) Program and Operations Management
      1. *Deputies, Senior Administrators, Researchers, and Program Evaluators*
   (3) District Management
      1. *Superintendent and School Board*
      2. *Legal*

By Educational Program

Second, each line item expenditure is assigned to one of six broad *program* categories found in most school districts: General Education, Special Education, Chapter 1/Title I, Bilingual/ESL, Vocational Education, and Other (categorical) Programs. Special Education, Chapter 1/Title 1, Bilingual/ESL, and Vocational Education may include expenditures related to both restricted fund sources (categorical), and from general tax levies—to arrive at the total cost of a program regardless of the source of revenue. Other Programs will include expenditures from only restricted sources of funds (categorical dollars) that are not related to Special Education, Chapter 1/Title 1, Bilingual/ESL, or Vocational Education. General Education is the remainder—that is, expenditures related to unrestricted funds from general tax levies that are not related to Special Education, Chapter 1/Title 1, Bilingual/ESL, Vocational Education, or Other Programs. For greater detail and more sophisticated analysis, each program category can track expenditures of up to 90 related subprograms. Also, expenditures related to programs can be traced by a school district's various funds (so that In$ite's *total view of expenditures by program*—including restricted and unrestricted funds—can be seen in contrast to a district's *traditional regulatory view of programmatic expenditures by restricted fund only*).

By Location—Site and Grade Level

Third, each district location code is associated with the central office or a specific school building, with individual sites identified according to the grade levels served or other nonschool purpose: Elementary School, Middle School, High School, Alternative School, Other Schools, or Non-School. Central costs benefiting students and teachers, such as transportation and food service, are then allocated to individual school sites, using an appropriate metric (e.g., square footage of building for school-building maintenance and utilities, the number of route miles by school for transportation, or the number of meals served by each school for allocating food service costs) to capture the level of resources each school receives. If necessary, employee benefits are likewise allocated to schools by detail function and program as a proportion of employees' salaries by job category (teachers may have different benefits from paraprofessionals or substitute teachers, for example), therefore, employee benefit expenditures follow the related salary expenditure. Other expenditures, such as all costs for capital, debt service, and pass-through payments to other agencies, are "non-allocated" to school sites and are labeled as "Other Commitments." By keeping "Other Commitments" at the district level, realistic comparisons between schools are possible.

In$ite's methodology overcomes the major shortcomings exhibited by previous research efforts, capturing all costs by function, program, and location. Equity may now be analyzed at the individual school level, exploring all resources used to benefit children at a particular site. This approach facilitates true school comparisons on an intra- and cross-district basis, allowing for precise analysis of resources by function, program, and school. Furthermore, by focusing on interschool differences within a single district or multiple districts within a state, rather than interdistrict differences alone within a state, In$ite facilitates more meaningful comparisons.

HORIZONTAL EQUITY

Equity is divided into two broad categories: *horizontal equity*, that is, funding all equal students equally; and *vertical equity*, based on the assumption that students with different needs and problems should rightly be treated differently. *Horizontal equity*, analyzed first, is based on the assumption that all equal children require similar funding, which raises the following questions:

For Total School-Site Resources:

1. Of the *total resources* spent at the school site, how equitable are their distribution *across all schools* within the district?
2. Of the *total resources* spent at the school site, how equitable is distribution across groups of *schools by type* (Elementary Schools, Middle Schools, and High Schools).

For School-Site Resources, by Function and Level:

1. How equitable is the distribution of *school-site resources* across all schools in the district *by In$ite function* (Instruction, Instructional Support, Operations, and Leadership)?
2. How equitable is the distribution of *school-site resources* across groups or types of schools *by education level* (Elementary, Middle, and High Schools) for each In$ite function?

For School-Site Resources by Educational Program:

1. How equitable is the distribution of *school site resources* across all schools in the district *by In$ite program*?
2. How equitable is the distribution of *school site resources* across groups of schools *by education level* [*Elementary Schools, Middle Schools, and High Schools*] *and by In$ite program*?

A number of statistical methods are used to determine the variation in per pupil expenditures across schools within a single district. Each statistic sheds light on the degree of equity achieved from a slightly different perspective.

1. *Mean:* Average per pupil expenditure.
2. *Median:* Central point of per pupil expenditure by school, weighted by student. In other words, the median per pupil expenditure is the expenditure level at the school where the middle sample student happens to be.
3. *Range:* Difference between the highest and lowest per pupil expenditures by school.
4. *Restricted Range:* Difference of the scores between 5th and 95th percentiles, eliminating the possible distortion caused by the 10 percent outliers.
5. *Coefficient of Variation:* Calculated using the standard deviation divided by the mean (the average of the data sample), shown as either a decimal or percent. It ranges from 0.0 to 1.0 or, as a percent, from 0 percent to 100 percent. A zero or 0 percent means that the object is uniformly distributed among all children by their unit of analysis (district, school, program). Assuming the data follows a normal distribution, an equitable standard score for the Coefficient of Variation is 10 or less percent (0.10).
6. *Gini Coefficient:* Based on economic measures of income inequality, the Gini coefficient indicates how far the distribution of per pupil objects is by plotting the percentage of students with an equal percentage of object based on the Lorenz curve. A perfect distribution would show a 45-degree line, with a perfect relationship between the distribution of pupils and the distribution of money. If the resources are not equitable, the Gini graph will be a concave line that falls below the 45-degree line. The smaller the Gini coefficient, the more equitable the distribution of the resources. While no exact standard or benchmark has been set for the Gini coefficient, generally it should be a value below 0.1 to suggest equity in spending.
7. *McLoone Index:* The McLoone index focuses attention on the bottom of the distribution, signaling the degree of inequity for those schools spending below the median. The McLoone index provides the *ratio* of the values of all observations below the median point to the corresponding sums if all observations below the median actually had

the value of the median. The index ranges from 0 to 1.0 with a 1.0 indicating perfect equality. In practice, most equalized district data indicate a range from .70 to .95. While a national norm has not been set for the McLoone index, a score of higher than .90 is thought desirable.

The District Analyzed

We illustrate the power of using three-dimensional analysis to shed light on equity concerns by applying the methodology to "Mayville Public Schools" (Mayville), a fictitious name but an actual American public school district with 41,231 pupils attending 81 schools, and total expenditures of $245.4 million in 1993-1994. The purpose of this case study is to illustrate how and why the school-site level can and should be the unit and focus of equity analyses. We use "Mayville's" data to illustrate the methodology, and further encourage research to determine benchmarks and best practices for resource allocation policy. Table 7.1 summarizes the district by the education level of schools.

A first step in understanding school-level equity is to determine the percentage of the district's total resources that reach the school site. The In$ite location dimension distinguishes between Mayville's Central Office costs, Other Commitments, and School-Site expenditures (including both direct expenditures and those allocated from the Central Office).

The non-allocated Central Office costs and the expenses for Other Commitments (expenditures for capital projects, debt service, and the like) comprise the resources that do not have a direct impact on students and schools in the current fiscal year. As displayed in Table 7.2, these central office expenditures total $23.539 million, which is 9.59 percent of total costs. Other Commitments total $17.290 million—7.05 percent of Mayville's total budget. Therefore, the funds remaining to be analyzed with respect to school-level equity total $204.534 million or 83.36 percent of Mayville's total budget.

Table 7.1. Schools and Enrollment by Type, Total, and Percent "Mayville Public Schools" 1993-1994

Type	*Number of Schools*	*Enrollment (Pupils)*	*Percent*
Elementary	59	23,008	55.80
Middle Schools	10	7,265	17.62
High Schools	5	10,237	24.83
Alternative Education	N/A	721	1.75%
Total	81	41,231	100.00%

Table 7.2. Central, Other Commitments, and School Site Costs by Dollars, Percent & Per Pupil "Mayville Public Schools" 1993-1994

	Total Dollars:	*Percent of Total*	*Per Pupil Costs*
1. Central Office:	$ 23,539,073	9.59 %	$ 571
2. Other Commitments:	17,290,301	7.05 %	419
3. School Sites:	204,533,876	83.36 %	4,961
Total:	$245,363,250	100.00%	$5,951

The per pupil figures for Central Office, Other Commitments, and School-Site expenditures are likewise important for comparison. Each of the total dollar amounts for these three categories is divided by Mayville's total of 41,231 pupils. Table 7.2 shows that the per pupil cost reaching the school ($4,961 per pupil), of total spending ($5,951 per pupil) is reduced by $990 (including $571 per pupil to pay for the Central Office and $419 per student for Other Commitments). The remaining figure for per pupil expenditure at the school site, $4,961, becomes the foundation of our school-specific equity analysis.

School-site equity involves dispersion analysis of total school-site expenditures of $204.5 million using the standard equity measures. If we assume that the $204.5 million is equally distributed among schools, the district-wide mean per pupil expenditure at the school site would be $4,961 or 83.36 percent, also shown in Table 7.2. Once resources reach the school, the next step is to analyze how the expenditures are used, and how much directly reaches and benefits students and their teachers in the classroom.

As shown in Table 7.3, of the $204.5 million reaching schools all together, some $125.9 million or 61.54 percent reaches the classroom for Instruction (including the cost of Face-to-Face Teaching and Classroom Materials). In per pupil dollars, $3,053 per pupil goes for Instruction, out of a total of *$4,961* per pupil that reaches the school sites. Taken as a percent of total per pupil district spending of *$5,951*, however, Instructional spending of $3,053 per pupil is 51.30 percent. In summary, the district spends $5,951 per pupil (100%) overall, of which $4,961 per pupil—or 83.36 percent—reaches schools as a group; and $3,053 per student or 61.54 percent of school-site spending and 51.30 percent of district-wide total per pupil spending are expended in classrooms for Instruction. Other school-site costs include 9.04 percent (of the district per pupil of $5,951) or $538 per pupil for Instructional Support, 18.48 percent or $1,100 per pupil for School-Site Operations, and 4.54 percent or $270 per student for Leadership at the schools.

Table 7.3. School-Site Expenditures for All Schools by Function "Mayville Public Schools" 1993-1994

Function/ Sub-function	*Expenditures*	*Expenditures per pupil*	*% of Total School-Site Expenditures ($4,961) per Pupil*	*% of Total District Expenditures ($5,951) per Pupil*
Instruction:	$125,871,164	$3,053	61.54%	51.30%
Face-to-Face Teaching	120,017,878	2,911	58.66%	48.92%
Classroom Materials	5,853,286	142	2.86%	2.39%
Instructional Support:	22,174,323	538	10.84%	9.04%
Pupil Support	13,144,177	319	6.43%	5.36%
Teacher Support	1,998,714	48	0.97%	0.81%
Program Support	7,031,432	171	3.44%	2.87%
Operations	45,341,234	1,100	22.17%	18.48%
Leadership	11,147,155	270	5.44%	4.54%
Totals	$204,533,876	$4,961	100.0%	83.36%

Horizontal Equity Analysis

The following analysis demonstrates how rigorous, disciplined school-site data can be used to examine equity. This analysis illustrates use of only a selected portion of the possible combinations of the school site data from In$ite. The views of data analyzed and discussed below include:

Analysis by School Level/Type

- Examines total expenditures for all schools.
- Dispersion: Total spending for all schools
- Equity of total spending by school type or level

Analysis by Function

- Examines expenditures by function.
- Equity of total school spending by function
- Equity of Instructional spending by school type/level
- Equity of Instructional Support spending by school type/level
- Equity of Operations spending by school type/level
- Equity of Leadership spending by school type/level

Analysis by Program

- Examines general education expenditures versus expenditures for all programs (for all schools) to isolate variabilities.
- Equity of Instructional spending by program
- Dispersion: General Education versus All Programs for all schools

- Dispersion: General Education versus All Programs for Elementary Schools
- Central Tendency for Middle Schools
- Dispersion: General Education versus All Programs for Middle Schools
- Central Tendency for High Schools
- Dispersion: General Education versus All Programs for High Schools

Analysis by School Type/Level

Dispersion: Total Spending for All Schools

This section analyzes the horizontal equity of total school-site spending for Elementary, Middle, and High Schools; and for all schools in aggregate (excluding central office costs). See Table 7.4, column 1 for measures of dispersion. For all schools in the aggregate the range is $7,045 per pupil, with a maximum school spending score of $9,022 per pupil and a minimum of $1,977. The restricted range, which eliminates the five percent at the extremes (top and bottom), is $2,596 per pupil.

The Coefficient of Variation (COV) for All Schools, measuring the distribution of expenditures by school against the district-wide mean, was 0.128 or 87.2 percent. A "desirable" COV, as suggested by Odden and Picus (1992), is 90 percent or more. The Gini coefficient, measuring the proportion of resources spent on the same proportion of schools, was 0.121, with the national benchmark being a Gini coefficient of 0.1. The McLoone index, showing the ratio of spending on schools *below* the median to the spending necessary for those same schools to be *at* the median, was 0.878. This McLoone statistic is slightly below the accepted equity standard of 0.9.

Overall, school-site expenditures for all functions in Mayville appear slightly less equitable than acceptable standards when all schools are grouped together. Before we move to an analysis of expenditures by function and program, we first apply these same equity measures of all school-site expenditures across groups of schools by type (Elementary, Middle, and High Schools.)

Equity of Total Spending by School Type/Level

When distribution of all school-site expenditures is analyzed for grade level of school, distinct patterns begin to emerge. As shown in Table 7.4, column 2, Elementary Schools spend the least overall—with a median of $4,753 per pupil and mean of $4,709 per pupil. Middle Schools come next in overall spending, with a median of $5,393 and mean of $5,175 per pupil; and High Schools rank the highest at $5,621 as a median and $5,147 per student for the mean. While these results are not entirely unexpected, they do illustrate

Table 7.4. Horizontal Equity Per Pupil School-Site Expenditures by School Type or Level "Mayville Public Schools" 1993-1994

Equity Measure	*All Schools*	*Elementary Schools*	*Middle Schools*	*High Schools*
1. Median	$4,825	$4,753	$5,393	$5,621
2. Mean	$4,961	$4,709	$5,175	$5,147
3. Maximum	$9,022	$9,022	$7,483	$6,066
4. Minimum	$1,977	$1,977	$4,500	$4,394
5. Range	$7,045	$7,045	$2,983	$1,672
6. Restricted Range	$2,596	$2,410	$841	$1,510
7. Coefficient of Variation	0.128	0.142	0.134	0.111
8. Gini Coefficient	0.121	0.076	0.105	0.062
9. McLoone Index	0.878	0.901	0.912	0.859

how misleading the district-wide mean of $4,961 per pupil for school-site costs can be (shown in Table 7.2, calculated by dividing all school-site expenditures by all pupils).

Even the mean of actual school-site expenditures, as shown in Table 7.4, does not indicate that the dispersion may fall into a systematic pattern by school. Even though Elementary Schools as a group spent the least on average of the three school types, they also had the highest and lowest outliers, indicating that the largest degree of inequity is within this school-type group. As shown in Table 7.4, row 5, Elementary Schools exhibit the widest range at $7,045 pupil, compared to the Middle School range of $2,983 per pupil and High Schools of $1,672 per student. The restricted range (eliminating the highest and lowest 5% of the distribution) narrows Elementary School dispersion to $2,410 per student, while Middle Schools are still less dispersed in total spending at $841 and High Schools at $1,510 per pupil restricted range differences.

Measures of equity indicate that resources are distributed somewhat more proportionately, with the Gini coefficient at 0.121 for all schools, 0.076 for Elementary, 0.105 for Middle, and 0.062 for High Schools. However, the Coefficient of Variation (COV) shows slightly greater variability for Elementary Schools spending than the other two groups: Elementary Schools is 0.142, Middle school is 0.134, and High school is 0.111, with the All-Function COV at 0.128. Again, while High Schools as a group spend the most with a mean of $5,255 per pupil, as a type they are more equitable than the other types. The McLoone index for the below the average spending schools by type shows an overall (All School) score of 0.878, Elementary Schools at 0.901, Middle Schools at 0.912, and High Schools at 0.859. This indicates that according to the McLoone indexes, the "low-spending" High Schools are less equitable than the below-the-median Middle and Elementary Schools.

Once the analysis moves from dispersion measures of *aggregate* expenditures to expenditures by *function* and educational *program*, we shall see more variation and greater dispersion for all schools as a group *and* for schools by grade-level type.

Analysis by Function

Equity of Total School Spending by Function

Our application of the standard equity measures will determine the extent to which functional spending is equitably distributed. Table 7.5 displays these measures for each of the four major functional areas using expenditures for all schools as a group. (Note that the omitted functional area, Other Commitments, eflects all district-wide costs such as debt service and payments to other school districts, thus not attributed to individual school sites.)

Interestingly, the range is widest for the two most visible and costly school-site functions: Instruction and Operations. The range for Instruction was $5,081 per pupil and Operations was $3,946 per pupil. The restricted range was considerably lower for both, at $1,273 and $1,098 per pupil, respectively, indicating significant outliers at the top and/or bottom of the distribution.

Analysis using additional equity measures provide more insight into the nature of the inequities highlighted earlier. The McLoone index, for example, as shown in Table 7.5, row 9, indicates greater equity for Instruction at 0.943 than for any of the other functions, with Instructional Support at 0.887, Operations at 0.804, and Leadership having the lowest ranking for equity at 0.799 with the overall measure at 0.878. The Coefficients of Variation were 0.129 (meaning that resources are distributed proportionately across 87.1% of the schools) for Instruction, 0.103 (89.7%) for Instructional Support, 0.310 (69.0%) for Operations, 0.221 (77.9%) for Leadership, and 0.128 (87.2%) for All Functions together.

Table 7.5. Horizontal Equity Per Pupil School-Site Expenditures for All Schools by Function "Mayville Public Schools" 1993-1994

Equity Measure	*All Functions*	*Classroom Instruction*	*Instructional Support*	*School Operations*	*School Leadership*
1. Median	$4,825	$2,921	$518	$1,076	$256
2. Mean	$4,961	$3,053	$538	$1,100	$270
3. Maximum	$9,022	$6,836	$1,098	$4,030	$645
4. Minimum	$1,977	$1,755	$138	$84	$146
5. Range	$7,045	$5,081	$960	$3,946	$499
6. Restricted Range	$2,596	$1,273	$393	$1,098	$225
7. Coefficient of Variation	0.128	0.129	0.103	0.310	0.221
8. Gini Coefficient	0.121	0.123	0.111	0.157	0.131
9. McLoone Index	0.878	0.943	0.887	0.804	0.799

The Gini coefficients were 0.123 for Instruction, 0.111 for Instructional Support, 0.157 for Operations, and 0.131 for Leadership, with Gini coefficients for all four school-site functions at 0.121. Thus, overall school-site equity was higher than equity evaluated using individual functions, indicating that the greater proportion of expenditures were distributed proportionately across 93 percent of the schools.

Resource dispersion is greater by functional expenditure than for overall school spending, demonstrating the potential of doing functional analysis by school using In$ite. As we compare the Coefficients of Variation, 87.1 percent for Instruction, 89.7 per percent for Instructional Support, 69.0 percent for Operations, 77.9 percent for Leadership, and 87.2 percent for All Functions together were dispersed proportionally. The evidence demonstrates that multifunctional analysis is superior to evaluating only one school-site expenditure variance such as *Instruction* (see Hertert, 1995).

Furthermore, Sherman (1996) noted similar tendencies in Texas and Ohio data: that the Coefficients of Variation for schools in Ohio, for example, was 48.8 percent overall, with the Coefficient of Variation being much greater for some functions: 58.0 percent for Instruction, 119.0 percent for Student Support Services, 141.0 percent for Instructional Staff Support Services, 73.1 percent for School Administration, and 34.9 percent for School Operations (see Sherman, 1996, p. 51). Thus, the Pelavin Institute study concluded:

> The variation in both expenditures per student and the share of expenditures for different school-level functions—as measured using the Coefficient of Variation—differed significantly from function to function. In both Ohio and Texas, total expenditures per student and expenditures per student for instruction showed the smallest amount of variation, while expenditures per student for support services for students and instructional staff showed the largest variation; expenditures per student for school administration fell between the two extremes in both states. Similar patterns of variation were found when the shares of total expenditures for these functions were examined: the smallest amount of variation was found for Instruction, the largest was found for Support Services for students and Instructional Staff, and school Administration fell between the other functions (p. 46).

Equity of Instruction Spending by School Type/Level

We are especially interested in the possibility of inequity within specific functions when expenditures are analyzed by grade levels. This in-depth examination of school-site expenditures will indicate whether the significant dispersion by function occurs *within* types of schools, or *between* types of schools for each function, starting with Instruction.

Table 7.6 presents the statistics for per pupil expenditures on direct *Instruction* for Elementary, Middle, and High School groups. The median and mean show differences in per pupil spending for Instruction across grade levels,

Table 7.6. Horizontal Equity Per Pupil of School-Site Expenditures for Classroom Instruction by School Type or Level "Mayville Public Schools" 1993-1994

	Instruction			
Equity Measure	*All Schools*	*Elementary Schools*	*Middle Schools*	*High Schools*
1. Median	$2,921	$2,864	$3,031	$3,261
2. Mean	$3,053	$2,898	$3,015	$3,259
3. Maximum	$6,836	$6,836	$3,782	$3,842
4. Minimum	$1,755	$1,755	$2,711	$2,846
5. Range	$5,081	$5,081	$1,071	$ 996
6. Restricted Range	$1,273	$1,218	$ 492	$ 895
7. Coefficient of Variation	0.129	0.135	0.084	0.101
8. Gini Coefficient	0.123	0.142	0.080	0.062
9. McLoone Index	0.943	0.841	0.943	0.933

indicating some degree of inequity across types of schools. Elementary Schools spent $2,864 per pupil at the median for Instruction, with $2,898 per pupil as the mean. Middle Schools spent slightly more than Elementary Schools on direct Instruction, with $3,031 per pupil as the median and $3,015 as the mean. By contrast, High Schools as a group spent the most on direct Instruction, with $3,261 per pupil as the median and $3,259 as the mean.

Despite the apparent inequity between types of schools, the inequity *within* types of schools is perhaps of greater concern. The systematic analysis of this issue is made much more convenient for a wide sample of districts and schools with the advantage of In$ite's methodology and technology. Focusing again on Mayville, the range of spending on direct Instruction within Elementary Schools was $5,081, with a high of $6,836 per pupil and a low of $1,755. The range within Middle Schools as a type was much smaller at only $1,071 per pupil, with a high of $3,782 per pupil and low of $2,711 per pupil. High schools, finally, spent the most dollars on Instruction overall (mean of $3,259 per pupil) and had the greatest degree of equity within the group. The range of per pupil expenditures for Instruction in High School was only $996 per pupil, with a high of $3,842 and a low of $2,846 per student. Instruction for All Schools was the same as Elementary since this category registered the greatest range, $5,081 per student.

The restricted range again shows the Elementary Schools having the greatest degree of inequity within school type. Elementary Schools had a restricted range of $1,218, nearly 25 percent higher than the restricted range of high schools (at $895), and more than double the restricted range of middle school expenditures (at $492) on Instruction.

The McLoone index for observations under the median indicates that Elementary School Instruction was below the equity benchmark of .090 at

0.841 while Middle (0.943) and High Schools (0.933) were above the norm, with All Schools together at 0.943. The Coefficient of Variation parallels this trend, with 86.5 percent for Elementary, 91.6 percent for Middle, and 89.9 percent for High Schools, with the All-School COV of 87.1 percent. The Gini coefficients also show the existence of greater inequities among the Elementary Schools than within the other groups: 0.142 for Elementary, 0.080 for Middle, and 0.062 for High Schools were distributed proportionately. Hence, the greatest dispersion of Instructional expenditures is reflected in these measures, even though Elementary Schools had highest spending at the "top" of the distribution ($6,838 per student) for Instruction and lowest at the bottom of the distribution at $1,755 per pupil.

Thus, while Elementary Schools spent less per pupil on direct Instruction on average, than did Middle Schools and High Schools, they evidenced far greater variation in the distribution of these expenditures. Elementary Schools consistently demonstrated a higher degree of inequity using several different measures, including the McLoone index, Gini coefficient, and COV. With all three types of schools subjected to this level of analysis, only Elementary Schools' expenditures per pupil on Instruction fell below generally accepted standards of equity.

Equity of Instructional Support Spending by School Type/Level

These expenditures, as shown in Table 7.7, reflect spending on Instructional Support, the cost of those services which typically take place outside of the regular classroom environment but still involve face-to-face interaction with students (e.g., guidance, library services, extracurricular support), or direct support of teachers and programs, such as curriculum development, staff development, and support for district programs. The analysis, as shown in Table 7.7, seems to indicate that Elementary Schools spent more on

Table 7.7. Horizontal Equity Per Pupil School-Site Expenditures for Instructional Support By School Type "Mayville Public Schools" 1993-1994

	Instructional Support			
Equity Measure	*All Schools*	*Elementary Schools*	*Middle Schools*	*High Schools*
1. Median	$518	$522	$468	$502
2. Mean	$538	$573	$470	$486
3. Maximum	$1,098	$1,098	$577	$612
4. Minimum	$138	$138	$355	$402
5. Range	$960	$960	$222	$210
6. Restricted Range	$393	$462	$99	$198
7. Coefficient of Variation	0.103	0.221	0.113	0.151
8. Gini Coefficient	0.111	0.118	0.056	0.159
9. McLoone Index	0.887	0.892	0.924	0.881

Instructional Support with a median of $522 and mean of $573 than Middle Schools with median of $468 and mean of $470 or High Schools with a median of $502 per pupil and a mean of $486.

Once again, we see that the degree of inequity *within* school types or levels is greatest for Elementary Schools. For per pupil expenditures on Instructional Support, the range within Elementary Schools is $960, with a high of $1,098 per pupil and a low of $138 per pupil, with a restricted range of $462. High Schools have a high or maximum score of $612 per pupil and a low of $402 per pupil for a range of $210 per student. Middle Schools demonstrate a somewhat similar distribution of resources for Instructional Support, with a range of $222 per pupil, and a maximum of $577 and a minimum of $355.

Spending on Instructional Support seems to contain a number of outliers, because the restricted range of spending within each school type is much less than the full range. With the top and bottom 5 percent of the school distribution eliminated, the restricted range for Elementary Schools drops to $462, although it is still much higher than the High School restricted range of $198 and the Middle School restricted range of only $99. The All-School restricted range, Table 7.7, column 1, is $393.

Indicators of equity confirm these trends: Middle Schools as a group had a McLoone index of 0.924 (above the benchmark of 0.90), while Elementary Schools at 0.892 and High Schools at 0.881 fell just below the standard, although not by much. The Gini coefficients also support this contention, with Middle Schools as a group being more equitable with other Middle Schools at 0.056, while Elementary Schools were 0.118 and High Schools at 0.159 and All Schools at 0.111

The COV found 88.7 percent of Middle Schools distributed around the mean, calculated as the Standard Deviation divided by the Mean, while Elementary Schools showed 77.9 percent, and High Schools at 84.9 percent. All Schools had a COV of 0.103 or 89.7 percent proportionately funded. Thus, Middle Schools spent less on Instructional Support as a function but had greater equity within the category; but Elementary Schools spent the most per pupil but again had the greatest dispersion, as the analysis shows.

Equity of Operations Spending by School Type/Level

Analysis of the *operations* function (including building-level expenditures for utilities, maintenance, food services and transportation) provides some surprising evidence about day-to-day costs for different types of schools. Conventional wisdom seems to suggest that high schools, with their more diversified and specialized facilities, would be the most expensive to operate. However, these data suggest that in Mayville, middle schools were in fact more costly to operate, perhaps because of their smaller size, complexity, and higher per-student costs, as compared to the larger high schools.

Table 7.8. Horizontal Equity Per Pupil School-Site Expenditures for Operations By School Type "Mayville Public Schools" 1993-1994

	Operations			
Equity Measure	*All Schools*	*Elementary Schools*	*Middle Schools*	*High Schools*
1. Median	$1,076	$1,018	$1,349	$1,066
2. Mean	$1,100	$1,010	$1,362	$1,085
3. Maximum	$4,030	$4,030	$2,939	$1,391
4. Minimum	$84	$84	$1,007	$906
5. Range	$3,946	$3,946	$1,932	$485
6. Restricted Range	$1,098	$947	$1,162	$429
7. Coefficient of Variation	0.310	0.327	0.297	0.156
8. Gini Coefficient	0.157	0.162	0.138	0.086
9. McLoone Index	0.804	0.763	0.897	0.944

As shown in Table 7.8, the 10 Middle Schools were the most costly to operate, with a median of $1,349 and a mean of $1,362. Next were the seven High Schools with $1,066 (median) and $1,085 per pupil as mean score. The 59 Elementary Schools were the least costly with a median of $1,018 per pupil and a mean of $1,010 per pupil. Yet, Elementary School cost dispersion was again the greatest, as illustrated by the range and equity calculations.

Operations as a category show the greatest variance among the In$ite functions, with a range for Elementary Schools of $3,946 per pupil (difference between a maximum of $4,030 and minimum of $84 per pupil) but a restricted range of much less at $947 per pupil as 10 percent of the outliers are excluded for Elementary Schools and $1,098 as the restricted range for All Schools. High Schools vary between $1,391 per pupil and $906, a range of $485 and a restricted range of $429. And Operations spending for All Schools has a median of $1,076 and a mean of $1,100 per student for the 81 schools in the system. The maximum-spending school overall is $4,030 and the minimum, $84 per pupil (the Elementary outliers).

A cautionary note is important here relative to the interpretation of operational spending. We have observed that many, if not most, of the districts subjected to financial analysis using In$ite have not done any meaningful work on the tracking and cost accounting of operational expenditures at the school level. In$ite, while providing the opportunity to use allocation methods for expenditures that are meaningful to the specific district under analysis, may "smooth out" the variance in what we believe would be actual spending patterns on Operations. Hence, costs may be allocated on a per pupil basis for Transportation and numbers of meals served for allocating Food Service expenditures, which comes closer to actual school costs but tends to reduce variability.

Given the variation we can already observe in operations using In$ite, we recommend a more sophisticated approach to capturing actual operational expenses and evaluating the efficiency in school plant operations. For example, installing electrical, gas, and water meters in each school would record the true use by school, rather than relying on an allocation metric such as square feet by building to divide up the utilities costs among sites.

As has been the pattern, Elementary Schools are the least equitable by every measure of dispersion: on the McLoone index, Elementary Schools have 0.763, Middle Schools 0.897, and High Schools 0.944. All Schools have a McLoone of 0.804. The COV likewise shows Elementary Schools at 0.327 (or 67.3% of the Elementary Schools are funded proportionately), Middle Schools at 0.297 (or 70.3%), and High Schools at 0.156 (or 84.4%), with All Schools having a COV of 0.310 (69%). The Gini coefficients are 0.162 for Elementary Schools, 0.138 for Middle Schools, and 0.086 for High Schools; All Schools had a Gini of 0.157. Thus, as was true of other functions, Elementary Schools showed the greatest inequities for building Operations.

Equity of Leadership Spending by School Type/Level

Finally, we examine school-site expenditures for the fourth function, school-site Leadership. Leadership expenditures include principals and assistant principals' salaries and benefits, as well as the costs associated with providing services in the school office (secretary, phone, copying). Table 7.9 presents the per pupil expenditures and analysis for the school Leadership for all schools and by school type or level.

The level of equity for the Leadership function is first shown by the district *median* of $256 per student for All Schools, with High Schools recording the highest median of $345 per pupil, Middle Schools next at $341, and Elementary

Table 7.9. Horizontal Equity Per Pupil School-Site Expenditures for Leadership By School Type "Mayville Public Schools" 1993-1994

	Leadership			
Equity Measure	*All Schools*	*Elementary Schools*	*Middle Schools*	*High Schools*
1. Median	$256	$226	$341	$345
2. Mean	$270	$228	$328	$317
3. Maximum	$645	$645	$473	$391
4. Minimum	$146	$146	$264	$236
5. Range	$499	$499	$209	$155
6. Restricted Range	$225	$201	$111	$141
7. Coefficient of Variation	0.221	0.246	0.218	0.176
8. Gini Coefficient	0.131	0.123	0.114	0.100
9. McLoone Index	0.799	0.875	0.811	0.828

Schools, $226. The *mean* scores were $328 per pupil for Middle Schools, $317 for High Schools, and $228 for Elementary Schools. The maximums and minimums show Elementary Schools to have the highest and lowest scores at $645 and $146 per pupil respectively, with a range of $449 and a restrictedrange of $201. Middle Schools have a maximum score of $473 and a minimum of $264, with a range of $209 and restricted range of $111 per pupil. High Schools, interestingly, have the lowest range at $155 per pupil and a restricted range of only $141, with a Maximum of $391 and Minimum of $236 per pupil. The All-School category data are identicla to Elementary Schools, which has the greatest outliers.

The McLoone Index mirrors these difference, with Elementary Schools highest at 0.875, High Schools next at 0.828, and Middle Schools at 0.811, with an overall McLoone for All Schools at 0.799. The Coefficent of Variation ranges from 0.246 (75.4%) for Elementary Schools, to 0.218 (78.2%) for Middle Schools, and 0.176 (82.4%) for High Schools. All Schools together have a Coefficent of Variation of 0.221 (77.9%). The Gini Coefficients go from 0.123 for Elementary, to 0.114 for Middle, to 0.100 for High Schools. Overall for Leadership, the Gini is 0.131. Thus, these data show the overall cost of Leadership to be highest at the High Schools but most dispersed sat Elementary Schools.

Analysis by Program

While functional analysis is helpful in understanding the distribution of resources across and within school types, students are also affected by resources for such programs as special education, bilingual/Limited-English-Proficient programs, Title I, vocational education, general education, and many others. Hence, a full analysis of the level of resources reaching children for their education should take into account both the functional *and* programmatic dimensions of school expenditures.

Our program analysis examines General Education expenditures as measured against combined cost of all program categories in schools: special education, bilingual/ESL, Chapter 1/Title 1, vocational education/school-to-work, other categorical programs, and general education. In other words, the subset of expenditures known as general education, compared to the grand total of all expenditures (including general education). (In$ite also provides the ability to evaluate up to 99 subprograms for each of the six umbrella programs listed above. For the purpose of this analysis we focus on the aggregate expenditures for the six programs.)

Expenditures on *Instruction* by program are most important since inequities may be related to the higher level of spending for special education, Chapter 1/Title I programs, and bilingual/ESL services, and other categorical programs in each school over general education for the "regular" student. The extra expenditures for these programs are likely to have their greatest impact on the functional expenditures for Instructional since other costs by function are more or less fixed (buildings, administration, food, libraries, etc.). Whereas, the greatest impact of Other Categorical programs is most likely to be observed in Instructional Support, because much of this restricted money goes toward staff development.

To gain a stronger understanding of how students with similar needs are being treated within each school type, we amplified the evaluation of programmatic spending with the comparison of expenditures for Instruction for *all* programs (General and Categorical) with expenditures for Instruction for general education *only*. We then recalculated the measures of equity to see to what degree expenditure differentials for direct Instruction were related to the presence or absence of special programs.

We hypothesize that in most cases, schools with large special-needs populations would spend more on Instruction than schools with few Title I, bilingual/LEP, or special education program students, thus increasing the absolute differences, the dispersion levels, and thus the inequities. However, in Mayville, our sample district, many expenditures for categorical programs are charged centrally by the district and then allocated (via In$ite) on a programmatic per pupil basis to schools. This per capita distribution has a "smoothing" or equalizing (rather than a dis-equalizing) effect on expenditures. While most schools in this district have students with some special needs, presumably the analysis would be more robust if the district were more precise about school-site accounting.

Equity of Instruction Spending by Program

Table 7.10 presents data on Instructional school-site spending by General Education alone, and by General and Special (categorical) programs together, for All Schools and Elementary Schools to illustrate the effects of program costs on school-site equity of Instructional expenditures. The medians for *General Funds*-only All Schools and *General Funds*-Elementary Schools for Instruction at \$2,540 and \$2,449, respectively (see Table 7.10, columns 1 and 3) are systematically *lower* than the *All Funds*-All Schools at \$2,921 and *All Funds*-Elementary Schools at \$2,864. This indicates that as expected the costs are higher when the categorical aid programs (Special Education, Title I, Bilingual Education) are included. The mean scores show similar patterns, with *General Funds*-All Schools averaging \$2,649 and *All Funds*-All Schools, \$3,028. Elementary *General* are lower at \$2,449 while *All Funds*-Elementary was higher at \$2,898 per pupil.

The range (row 5) for All Schools-*General* Education costs alone is lower at \$2,742 as compared to *All Funds*-All Schools at \$5,083 per pupil, with the restricted range for All Schools-*General* Funds at \$1,525 and *All Funds*-

Table 7.10. Horizontal Equity Per Pupil School-Site Expenditures for Instruction on All Programs v. General Education Only "Mayville Public Schools" 1993-1994

	Instruction			
	All Schools		*Elementary Schools*	
Equity Measure	*All Schools General Education Only*	*All Programs*	*Elementary Schools General Education Only*	*All Programs*
1. Median	$2,540	$2,921	$2,449	$2,864
2. Mean	$2,649	$3,053	$2,449	$2,898
3. Maximum	$3,850	$6,838	$3,537	$6,838
4. Minimum	$1,108	$1,755	$1,757	$1,906
5. Range	$2,742	$5,083	$1,780	$4,932
6. Restricted Range	$1,525	$1,273	$969	$1,281
7. Coefficient of Variation	0.114	0.129	0.135	0.117
8. Gini Coefficient	0.123	0.113	0.115	0.101
9. McLoone Index	0.933	0.942	0.913	0.948

All Schools, at $1,273. For Elementary Schools-General Education, the range goes from a maximum of $3,537 per student to a minimum of $1,757, for a difference (range) of $1,780 per pupil and a restricted range of $969 per student. With all programs for Elementary Schools included the range is much greater, with the maximum score at $6,838 and the minimum at $1,906 for a range of $4,932 (compared to just $1,780 for General Education alone) and a restricted range for All-Program Funds/Elementary Schools of $1,281 per pupil for Instruction.

Dispersion: General Education versus All Programs for All Schools

As shown in Table 7.10, the McLoone index was 0.933 for All Schools-*General Education* alone, and 0.942 for *All Schools*-All Program costs (General and Special). The COV is 0.114 (88.6%) for *All Schools*-General Education and 0.129 (87.1%) for *All Schools*-All Programs, indicating the effects on the dispersion of including Per Capital Special/Categorical costs on school-site equity. The Gini coefficient for General Education Instruction for All Schools is 0.123, and for All Costs-All School types (General and Categorical spending) was 0.113. Again, interestingly, the effects of removing all categorical funds is to make the schools appear more equitable since many special program expenditures (Special Education, Bilingual Education, Title I and II) are allocated to schools on a per pupil basis, acting as an equalizing agent.

Dispersion: General Education versus All Programs for Elementary Schools

These schools have a McLoone index of 0.913 for General Education versus 0.948 for All Funds—again showing the equalizing effects of special education services. The Gini coefficient for Elementary Schools-General Education alone for Instruction is 0.115 and for All Programs-Elementary School Instructional costs at 0.101, indicating greater equity because of the equalizing effects of per capita distribution of special program expenditures. The COV for Elementary Schools for General Education alone is 0.135 (86.5%) and 0.117 (89.9%) when all Special and General Education program dollars are included.

Central Tendency for Middle Schools

As shown in Table 7.11, Middle School General Education has a median score of $3,031 per pupil while the All Cost for Middle Schools is $3,092, with a mean of $2,612 for General Education and $3,015 per pupil for all Middle School expenditures for Instruction. The Maximum score for General funds is $6,059 per pupil and the Minimum, at $2,670, creating a range of $3,389 and a restricted range of $2,286. For All Cost of Instruction for Middle Schools, the Maximum is $3,782 and Minimum of $2,711, with a range of $1,071 and a restricted range of $492 per pupil.

Dispersion: General Education versus All Programs for Middle Schools

As shown in Table 7.11 the McLoone for Middle School equity is 0.904 and for All Costs is 0.943; the COV is 0.136 for the General Fund expenditures and 0.083 for All Funds. And the Gini coefficient for Middle School Instruction for General Education is 0.084 and for All Program Funds is 0.043. These data indicate again the equalizing effects of distributing special funds on a per pupil basis to schools since in every case the All Funds are more equitable than the General (non-Categorical) alone.

Central Tendency for High Schools

As shown in Table 7.11, columns 3 and 4, the *General Education*-High School median is $3,118 and the mean is $2,905 per pupil, while the median for *All Cost* for High School Instruction is $3,261 and mean is $3,259 per pupil. The Maximum for General Education High School Instruction is $3,850 and Minimum of $2,742 per student, with a range of $1,108 and restricted range of $628. Full Costs for High School Instruction, including Categorical and General Education, has a range of $996 (Maximum of $3,842, and Minimum of $2,846 per student) and restricted range at $895 per pupil for All Education-High School Instruction.

Table 7.11. Horizontal Equity Per Pupil School-Site Expenditures for Instruction on All Programs versus General Education Only "Mayville Public Schools" 1993-1994

	Instruction			
Equity Measure	*Middle Schools General Education Only*	*All Programs*	*High Schools General Education Only*	*All Programs*
1. Median	$3,031	$3,092	$3,118	$3,261
2. Mean	$2,612	$3,015	$2,905	$3,259
3. Maximum	$6,059	$3,782	$3,850	$3,842
4. Minimum	$2,670	$2,711	$2,742	$2,846
5. Range	$3,389	$1,071	$1,108	$996
6. Restricted Range	$2,286	$492	$628	$895
7. Coefficient of Variation	0.136	0.083	0.126	0.117
8. Gini Coefficient	0.084	0.043	0.053	0.101
9. McLoone Index	0.904	0.943	0.913	0.925

Dispersion: General Education versus All Programs for High Schools

The seven high schools showed a McLoone index of 0.913 for General Fund dispersion by school, and 0.925 for All Funds. The COV was 0.126 (87.4%) for General to 0.117 (88.3%) for All Funds. And the Gini coefficient was 0.053 for General Funds and 0.101 for All costs. All these findings show the equalizing effect of distributing expenditures on a per pupil basis, rather than on an actual school-site costs, requiring that the district's general ledger provide school-site specific data for each school and program.

VERTICAL EQUITY

Vertical equity recognizes the financial impact of differing needs of students. These needs are categorized according to educational programs: Special Education, Chapter 1/Title 1, Bilingual/ESL, Vocational, and General Education. Horizontal equity analysis examines how equitably a program's education dollars are expended *across all schools*; whereas, vertical equity examines how equitably dollars are expended vertically *among programs.*

"The Unequal Treatment of Unequals"

Thus far, we have examined levels of *horizontal equity* among schools, types, functions, and programs. That is, we tested spending for "the equal treatment of equals," giving all students the same weight in calculating per pupil costs. To test the In$ite data more fully, we now apply analysis on another real but disguised American district, the "Westville Public Schools" (Westville), to the notions of *vertical* equity: that is, the "unequal treatment of unequals." This approach often means spending more money on those students with greater intellectual, physical, or emotional needs or challenges. Also, some students may require more expensive training, such as vocational and technical preparation, which is regulated by state and federal policies limiting class size to ensure greater safety around mechanical and potentially dangerous equipment (see regulations under the Vocational Occupation Safety and Health Act, VOSHA, for example).

The difficulties related to vertical equity analysis seem to be three: (1) how to weight the differences among unequal groups; (2) how to determine the population of each unequal category once the characteristics of each vertical group are established; and (3) how to norm school-site weights, whether: against each school's General Education expenditures (Self-Norms), against District norms for General Education, or against some agreed on national weights for Special Education (Full and Part Time), Bilingual/ESL, Title I, and other programs, as against the cost per pupil of General Education. The problem then is answering the obvious questions: just how "unequal" are these unequal groups, compared to what norms or standards, created and approved (sanctioned) where and by whom?

Berne and Stiefel (1984) explain that "the selection of the unequal groups of pupils is a value judgment that must be made regardless of the way vertical equity is to be measured. Furthermore, if vertical equity is measured with weighted dispersion measures, the appropriate unequal treatment for the unequal groups must be specified, usually in the form of pupil weightings" (1984, p. 223). Table 7.12 does just that: it specifies preestablished weights for a range of "unequal" students or pupils requiring special settings (Homebound and Hospitalized), although the "handicapping conditions" which are weighted sound a bit outdated.

Table 7.12. Vertical Equity Weights for Students with Various Challenges

Title of Condition	*Vertical Weights*
1. Educable Mentally Impaired	1.50
2. Trainable Mentally Impaired	2.00
3. Severely Mentally Impaired	2.00
4. Emotionally Impaired	2.70
5. Learning Disabled	1.50
6. Hearing Impaired	1.50
7. Visually Impaired	5.01
8. Physically and Health Impaired	3.00
9. Severely Multiply Impaired	1.29
10. Homebound and Hospitalized	1.60

Source: (Berne & Stiefel, 1984, p. 228).

Problems in Determining Vertical Equity

It may be difficult, given the complex and changing nomenclature in the field, to determine just how many categories of "unequal" students there are in a district or school? How much more would (should) it cost to educate a physically or academically challenged student than a regular student? And how equitably are students within the same category treated across schools, levels, functions, and programs?

Two methods can be used for determining weights: First, a preset weighting scheme can be created and agreed on, and then imposed on the data in each school district or school. William T. Hartman (1981) was among the first to attempt that method through the National Educational Finance Project. Since our goal in this chapter is to examine spending equity among unequal groups in each school, then simply preestablishing ratios for whole states may have limitations since the district is the universe, within which each school is fiscally embedded.

Second, a school-district benchmark for Instruction can be established, based on the ratio of overall costs for regular and challenged students—which can then be applied to each school to get a fix on just how dispersed the school is from the district "norms." For example, let's say that regular education per pupil expenditures for Instruction are $4,000, Part-Time Special Education are $6,000, and Full-Time Special Education are $10,000. Then, the district-wide weighting for General Education at $4,000 is 1.0; the Part-Time Special Education weighting would be 1.50; and Full-Time Special Education at $10,000 against the baseline of $4,000 would be 2.50. These weights, then, become the basis for calculating the equity levels among groups by school, program, and function.

Third, other methods for measuring vertical equity are possible, including the regression-based relationship measures (wherein costs for special students are regressed against the percent of students in that category) and ratio measures. The latter can be performed by computing the ratio of average cost of per pupil expenditures for special students to the average cost per pupil for students NOT in the special or challenged category. This could be performed by assigning each student in the district the object level of his or her school. When the population of special needs groups is determined for each school site, then an average level of cost can be calculated for each special and regular group. A measure can then be computed as the ratio of the average per pupil costs for special groups to the average school-site costs for regular students not in these groups.

Determining Populations for Each Category

School districts sometimes have difficulty specifying the number of students that fall into each "unequal" category since students may be both bilingual and academically challenged, and/or Title I eligible; and the classification of students can be a long, tedious, and rather idiosyncratic process. Furthermore, it may be difficult to separate costs factors that are affected by special populations from those factors that are not. Many districts, for example, do not charge Leadership (e.g., superintendent and principals' salaries and benefits) and basic school Operations (maintenance, utilities, and food) to special student populations, whereas the expenditures for Instruction and Pupil Support may be powerfully affected by the scope and extent of a district's handicapped and challenged student population.

Demonstrating a Vertical Equity Model Using In$ite Data

Although little agreement exists on both the weighting of all types of students and the exact population of each category, this section demonstrates how vertical equity might be determined for all schools within a district using In$ite school-specific data, rather than district data across a state (as has usually been the case). From data in the Westville, we selected a sample school—fictitiously named "Edgewater Elementary School" (Edgewater Elementary)—applied weights, and then divided total adjusted pupil counts into total school-site spending on Instruction. In gross terms, we recalculated the "pupil counts" based on some metric concerning the ratio of regular education to categorical.

Furthermore, schools may not keep records on the treatment of each of these many subgroups of special students but may use broader categories instead, may place a mix of students into the same classroom, or may mainstream many or all of these students. Unless a district is quite large and thus has separate classrooms for various subgroups, it may be difficult to track costs to each school and classroom. We found when applying In$ite that we could figure out how many students are served in Special Education/Full Time, Special Education/Part Time, Bilingual/ESL, Title I and II, and Other Programs by school. We did not come close to the 10 or so "handicapping conditions" spelled out by Berne and Stiefel (1984; see Table 7.12 above) since the general ledgers of districts do not usually apply dollars to such fixed categories.

Analysis of Edgewater Elementary permits an exploration and application of a weighting system to costs at a school. These data can be compared to the overall district-wide costs for these subcategories of special students, as well as against the General Education/Categorical Education ratios for All Elementary Schools and particularly the Edgewater Elementary school itself. Westville has 34,143 pupils and spent $223,575,777 in 1994-1995, for a total per pupil expenditure of $6,548. In general, of this amount, $120,839,404 was expended on direct classroom Instruction, which is the focus of this section. We seek to learn the ratios of General Education (Instruction) costs for the regular education student to the expenditures for Full-Time Special Education, Part-Time Special Education, Bilingual/ESL, and Title I and II.

Table 7.13. Vertical Equity Spending on Classroom Instruction Per Pupil by School Type or Level as Percent of District Expenditure Total ($6,548 per pupil) "Westville Public Schools" 1994-1995

School Level or Type	*Per Pupil Expenditures for Instruction*	*Percent of District Total Per Pupil Costs ($6,548)*
Elementary Schools	$3,572	54.55%
Middle Schools	$3,171	48.43%
High Schools	$3,684	56.26%
All Schools	$3,539	54.05%

Instruction, the core element of education, varies across types or levels of school. As shown in Table 7.13, the Elementary Schools as a type spent $3,572 per pupil or 54.55 percent on Instruction, of the District's total cost per pupil of $6,548. Middle Schools expended less at $3,171 per student or 48.43 percent on Instruction, and High Schools expended $3,684 per student or 56.26 percent, while the District total cost for Instruction at all levels of schools was $3,539 per student or 54.05 percent reaching the classroom in the Westville.

These figures include expenditures for all programs. To investigate *vertical* equity for Instruction, we must separate all costs for General Education (called the General Education Marker, or GEM) and by each Program for Instruction, at the Total District Level, at the Elementary School Level, and finally by the specific school, Edgewater Elementary. This three-level analysis (District Instruction, Elementary School Instruction, and Edgewater Elementary Instruction) gives us three GEMs (General Education Markers) against which to compare each of the program Instruction expenditures in order to test the degree of vertical equity recorded in Westville. We can also compare these ratios to national markers as well, as Table 7.14 provides.

General Education versus Special Program Costs for Instruction

All Schools

As shown in Table 7.14, the General Education cost for Instruction for all Westville schools is $3,025 per pupil, which becomes the General Education Marker (GEM), as well as the baseline costs for Part-Time Special Education, Bilingual/ESL, and Chapter I. This "base" cost concept rests on the notion that Part-Time Special Education, Bilingual/ESL, and Chapter I/II pupils all receive the basic program PLUS supplemental expenditures under each categorical program. Full-Time Special Education students, however, are often separated by classroom or even by school or sub-school and thus receive their own (self-contained) amount, as shown in column 2 of Table 7.14.

Table 7.14. Vertical Equity Ratio of General Education to Categorical Programs Per Pupil Costs for All Schools—Instruction "Westville Public Schools" 1994-1995

Cost Center	*General Education Marker*	*Special Education Full Time*	*Special Education Part Time*	*Bilingual ESL Program*	*Title I and II Program*
Instruction Base Cost	$3,025	$0	$3,025	$3,025	$3,025
Program Cost	$0	$9,064	$1,879	$1,719	$1,542
Total Program Cost	$3,025	$9,064	$4,904	$4,744	$4,567
Vertical Equity Weight (VEW)	1.00	2.99	1.62	1.57	1.51

Table 7.14 also presents the cost of Full-Time Special Education for students in All Schools in Westville, at a total per pupil of $9,064. When taken against the General Education Marker (GEM) of $3,025, the ratio or Vertical Equity Weight (VEW) is 2.99, meaning that it is three times as expensive to educate a Special Education student full time than a general education pupil—a triple unequal treatment of unequals. This weight corresponds closely to those cited earlier by Berne and Stiefel, as shown in Table 7.12.

All Elementary Schools

Table 7.15 presents much the same analysis for expenditure comparisons for *Instruction* in All Elementary schools in the Westville Public Schools. This datum may provide a different benchmark for Edgewater Elementary, our case example.

The GEM (see column 1), weighted as 1.0, provides a baseline for calculating the level of unequal treatment of unequals, with Full-Time Special Education at $9,064 per pupil (the same as the district-wide costs since Westville has all Full-Time Special Education in a single special school).

Table 7.15. Vertical Equity Ratio of General Education to Categorical Programs Per Pupil Costs All Elementary Schools—Instruction "Westville Public Schools" 1994-1995

Cost Center	*General Education Marker*	*Special Education Full Time*	*Special Education Part Time*	*Bilingual ESL Program*	*Title I and II Program*
Instruction Base Cost	$3,040	$0	$3,040	$3,040	$3,040
Program Cost	$0	$9,064	$1,919	$1,579	$1,540
Total Program Cost	$3,040	$9,064	$4,959	$4,619	$4,580
Vertical Equity Weight (VEW)	1.00	2.99	1.63	1.52	1.51

The weight, then, is 2.99, close to the 3.0 level found district-wide in Table 7.12. Part-Time Special Education costs are the base of $3,040 plus the supplementary costs of $1,919 per pupil, for a total of $4,959—with a Vertical Equity Weight or VEW of 1.63. Bilingual/ESL adds $1,579 per pupil to the base or GEM of $3,040 for a total of $4,619 per pupil and a weight of 1.52. And Title I and II totaled $4,580 per student ($1,540 plus the GEM of $3,040) for a weight of 1.51.

Edgewater Elementary is a real and useful case study with its robust information on spending by program, thus providing an opportunity to test the Vertical Equity Weights (VEWs) against an internal school VEW, All-School VEW, and an Elementary School VEW. Edgewater Elementary has a total of 392 pupils, against whom the district-wide proportion of Full-Time Special Education pupils can be charged, four bilingual students, 138 Title I and II students, and 86 Part-Time Special Education students. The total school-site spending at Edgewater Elementary in 1993-1994 was $2,683,536 on 392 total pupils for a per pupil cost of $6,846, of which $1,833,127 or $4,676 per pupil (68.3%) reaches the classroom for direct Instruction.

Table 7.16 provides data on the special program costs against which the General Education Marker at Edgewater Elementary can be compared, before measuring these per pupil costs against the district-wide All School and All Elementary GEMs. Internally, as shown in column 1, the GEM at Edgewater Elementary for Instruction was $3,481 per pupil, which is higher for General Education than the district-wide average and the Elementary average Instructional expenditures. However, for internal norming purposes, the $3,481 is given a VEW of 1.0, against which the various special ("unequal") programs can be measured.

Special Education/Full Time at Edgewater Elementary cost $10,461 per pupil for Instruction, which provides a VEW of 3.01 ($10,461 divided by the GEM of $3,481 per pupil). The Part-Time/Special Education is the result of

Table 7.16. Vertical Equity Ratio of General Education to Categorical Programs Per Pupil Costs "Edgewater Elementary School"—Instruction "Westville Public Schools" 1993-1994

Cost Center	*General Education Marker*	*Special Education Full Time*	*Special Education Part Time*	*Bilingual ESL Program*	*Title I and II Program*
Instruction Base Cost	$3,481	$0	$3,481	$3,481	$3,481
Program Cost	$0	$10,461	$1,898	$2,510	$1,677
Total Program Cost	$3,481	$10,461	$5,379	$5,991	$5,158
Vertical Equity Weight (VEW)	1.00	3.01	1.55	1.72	1.48

combined cost of the base expense of $3,481, plus the supplemental cost of the program at $1,898, for a total Part-Time/Special Education expense of $5,379 per student. When compared to the General Education Marker, it provides a VEW of 1.55. Bilingual/ESL add $2,510 per pupil to the General Education base for a total of $5,991 and a VEW of 1.72. And Title I and II totals $5,158 per pupil ($3,481 plus $1,677) for a VEW of 1.48. Thus, when normed against itself, the Edgewater Elementary is very much in keeping with other VEWs except for Bilingual/ESL which runs more and has a higher VEW.

Multi-Vertical Equity Comparisons

Since vertical equity analysis by individual schools for Instruction is still in its infancy, we cannot readily determine which VEW to use: whether one based on a School-Specific VEW, a District-wide VEW, or one for each school level or type (e.g., an Elementary Level VEW). Thus, Table 7.17 compares the VEWs for Edgewater Elementary against (column 1) All Elementary Schools General Education Marker ($3,040 per pupil for Instruction), against (column 2) Edgewater Elementary itself ($3,481), and against the cost of Instruction for General Education at All Schools (column 3) in Westville ($3,025 per pupil). Down the far left side of Table 7.17 are the per pupil Instructional costs of the five programs at Edgewater Elementary: row 1 shows General Education at $3,481 per pupil, row 2 is Full-Time Special Education: $10,461; row 3 is Part-Time Special Education $5,379; row 4 is Bilingual/ESL at $5,991 per pupil; and row 5 is Title I and II at $5,158.

General Education

Row 1 of Table 7.17 shows the base or General Education Marker as compared to Edgewater Elementary costs: when compared to All Elementary, it registers a 1.15 to All Schools in Westville, meaning that at the General Education level, this school is comparable to other schools, establishing this school as a typical and useful case study.

Full-Time Special Education

Row 2 compares the VEW at Edgewater Elementary against All Elementary at 3.44, against its own marker at 3.01, and against all Westville schools at 3.46. Thus, this school does expend unequally on unequal students, even greater than the district or level average, by a factor of 3.01 to 3.46.

Table 7.17. Vertical Equity Weights (VEWs) Summary: "Edgewater Elementary School" Weightings by Expenditures Per Pupil for Instruction by Program Against (1) All-Elementary Schools, (2) Edgewater Elementary (Self), and (3) All Schools "Westville Public Schools" 1993-1994

Program Categories	*Spending Per Pupil (Edgewater Elementary)*	*1. All Elementary Schools ($3,040 per pupil)*	*2. Self (Edgewater Elementary) ($3,481 per pupil)*	*3. All Westville Public Schools ($3,025 per pupil)*
General Education	$3,481	1.15	1.00	1.15
Full-Time Special Education	$10,461	3.44	3.01	3.46
Part-Time Special Education	$5,379	1.77	1.55	1.78
Bilingual/ESL	$5,991	1.97	1.72	1.98
Chapter I and II	$5,158	1.70	1.48	1.71

Bilingual/ESL

Row 4 shows that the school's cost per pupil was $5,991, which is significantly greater than the district's All School and All Elementary school spending levels ($4,744 and $4,619 per pupil, respectively). Perhaps this cost level is related to the exceptionally small number of Bilingual/ESL students in this school, only four, which makes this more expensive.

Part-Time Special Education

Row 3 shows the Edgewater Elementary cost of $5,379 per pupil against All Elementary for a VEW of 1.77, against its own GEM for 1.55, and All Schools in the district for 1.78. Again, this indicates that these special students are receiving more resources, almost half again as compared to the baseline within the school and three-quarters more against All Schools and All Elementary schools.

Titles I and II

As shown in Table 7.17, row 5, this program at Edgewater Elementary cost $5,158 per pupil, which provides a VEW of 1.70 for All Elementary and 1.71 for All Schools in the district, but a 1.48 at Edgewater Elementary. These VEWs compare closely with Part-Time Special Education, which is interesting in and of itself.

Summary

In the absence of norms across more schools, districts, and states, we can only speculate about levels of unequal funding by program. Further, research is possible using In$ite, as we are able to compare both horizontally using dispersion analysis and vertically using various weighting systems. And it seems quite obvious that school-site analysis provides a treasure throve of information for testing the equity of American education, both among equals and among unequals.

Hence, we can get some sense of dispersion for Instructional Support, Operations, and Leadership, as well as Instruction, to give a clear overall picture of spending equity. We also were able to include and exclude categorical program costs, to see the effects of Special Education, Bilingual/ESL, and Title I and II on equity at each school and for each school level (elementary, middle, and high schools).

POLICY AND LEGAL IMPLICATIONS

The quest for intra-district fiscal equity is not an abstract goal, disconnected from the aspiration for instructional quality. Inequalities may exist for legitimate reasons, such as the need for compensatory programs, differential age and use of buildings, and other factors requiring the targeting of limited funds to one particular school rather than another.

But clearly, fiscal inequities can, and sometimes do, arise not due to appropriate disparities among students or facilities, but because of unequal political power, poverty, or other invidious distinction (see Linda Hertert's chapter in this volume). The appeal to the judicial process, when political pressure is unavailing or unavailable, is time-honored and noteworthy in the context of interdistrict equity suits. (See, e.g. *Campaign for Fiscal Equity,* 86 N.Y. 2d 307 (1995); *Robinson* v. *Cahill,* 62 N.J. 473 (1973), cert. den. 414 U.S. 976 (1973).

The depth of equity analysis has been greatly advanced by the recent models of school-site expenditure data in three dimensions (functions, programs, and school types.) With new reliable information becoming available—as states, districts, and the National Center for Education Statistics move to gather data by school—the capability to examine school-site expenditures has enormous implications for the future course of equity litigation, and resource allocation policy.

In great contrast to the volume of case law around the issue of funding disparities *between* school districts, funding disparities *within* school districts have been largely free of judicial scrutiny. If those pursuing equity in the courts can link the lack of school resources to the race and poverty of the children in those schools, then the court will likely take note. And in those state constitutions that guarantee equity in spending, the financial disparity data may be sufficient on its own.

One obvious reason for the absence of legal scrutiny has been a dearth of available data. In his current analysis of early interdistrict litigation, Michael W. LaMorte ascribes the failure of district plaintiffs in *McInnis* v. *Shapiro*, 293 F. Supp. 327 (Ill. 1968); aff'd sub nom. *McInnis* v. *Ogilvie*, 394 U.S. 322 (1969); and *Burrus* v. *Wilkerson* 310 F. Supp, 572 (Va. 1969) to the lack of "judicially manageable standards" for evaluating plaintiffs' contention that the educational needs of students were not being met, and also that courts lacked the expertise to devise an equitable funding plan. LaMorte explains:

> Questions raised in these early cases revealed the lack of empirical data necessary to have a clear and accurate understanding of the effects of a state's finance system. Additionally, these cases revealed other methodological shortcomings, such as the lack of consensus over the goals of schooling and especially what constitutes a "basic" education (1996, p. 343).

Clearly, the recent availability of school-site information has real implications. Until the last three years, limitations in technology and in district management information systems allowed courts to point to a lack of empirical data as a rationale for not reaching a decision. With recent improvements in the performance of personal computing hardware, and the advent of software products like In$ite, courts will soon know that today any school district in America is only a few thousand dollars and two to eight weeks away from having definitive school site expenditure information that can be measured against "judicially manageable standards." This section examines the implications of school-site inequalities for school law and policies. In particular, what legal precedents exist for treating school spending as the point of determining equity and thus the legality of public education funding structures?

Implications from the Perspective of the Case Studies

We now use the horizontal and vertical equity findings from the case studies of two districts, Mayville, and Westville public schools, presented in the previous section, to explore the policy and legal implications of school-site expenditure data. In part, the approach to legal analysis will depend on the extent of inequalities between various levels in the school organization. To date, data were mainly available between districts, starting with *Serrano*. Now, however, new information at the school level may change the legal arguments, as follows:

1. Inequities can be demonstrated between schools within a single district. Detailed examination of school-site expenditures for different functions and programs indicates discernible inequities *within* school districts, as captured using several traditional equity measures. The dispersion of all school-site expenditures per pupil in Mayville failed to meet generally accepted standards of equity using the COV, the Gini coefficient, and the McLoone index.

2. Discernible patterns of inequity within school districts emerge by type of school, with elementary schools receiving less than middle or high schools. Overall, elementary schools spent less per pupil at the school site than did middle schools and high schools. Of even more concern is the evidence that elementary schools also spent the least per pupil on instruction, as a percent of overall district per pupil expenditures. This finding refutes conventional assumptions that greater expenditures in the upper grade levels are largely for support services and school building operations.

3. Additional analysis suggests marked inequities within types of schools, especially within the elementary school group. According to several measures, dispersion is more pronounced within elementary schools than within middle schools or high schools. This may, in part, be a factor of sample size, since Mayville had many more elementary than middle and high schools. But, the extremes are great enough within the elementary school type or level alone to call funding equity into question.

4. Special education costs greatly increase the inequities in instructional spending by school. We analyzed school data with and without categorical aid, such as special education, bilingual/ESL, and Title I. The results are quite interesting both ways. First, with special education and other categorical included, we see just how much effect these large costs have on average school-site spending levels. Second, for purposes of comparability, we excluded everything but general education for the "regular" student. Even here, the horizontal inequalities are significant and may be the basis for legal action.

Legal Reasoning

Judicial scrutiny is not triggered by constitutional mandates of cross-district equity per se, but rather by equity and adequacy of states' instructional support (see *Campaign for Fiscal Equity, supra; Robinson* v. *Cahill*, 303 A.2d 273 N.J. 1973). Thus, courts would certainly want to know just how inequitable the funding of classroom Instruction is, in the same way the *Serrano*-style judicial rulings would be concerned if one District spent $6,000 per student and another $12,000 per student within the same state.

Greater spending by school and classroom, within districts, may in fact have a more significant effect on student learning and other results than interdistrict differences previously considered. Research seems to show that the closer the resources are to students, the greater their effect. Hence, the whole interdistrict approach so long used in school finance litigation falls short of comparing the important "units of analysis," the classrooms and school sites. One study found that increasing resources to the classroom in New York City would raise test scores significantly (see Cooper & Associates, 1994). Thus, it is important to consider how new evidence of intra-district inequity fits existing case law on school finance and what future directions may take shape.

We find the key legal precedent for reviewing intra-district inequity in the companion cases of *Hobson* v. *Hansen* (269 F. Supp. 401, 1967) and *Smuck* v. *Hobson*, 269 F. Supp 401 (D.D.C. 1967). While school-site expenditure data were neither widely available nor even considered at that time, Judge J. Skelly Wright scrutinized expenditures at the *school site* to determine the degree to which children in the Washington, DC Public Schools were receiving equal educational resources.

Interestingly, both *Hobson* and *Smuck* pursued equity not under the rubric of equal financial spending per se but rather as an issue of just how funding contributes to racial segregation and discrimination. Judge Wright found that by allowing considerable differences in the per pupil spending levels between the schools attended by children of different races, the DC superintendent and school board had "unconstitutionally deprived Negro and poor public school children of their right to equal education opportunity with white and more affluent school children" (1967, p. 404).

Much of Judge Wright's dictum in *Hobson I* and *Hobson II* would not survive judicial scrutiny today. For example, the court's ruling that the Constitution bars inequities between schools attended separately by rich and poor pupils, 269 F. Supp. at 496 ff., was subsequently addressed in the negative by the U.S. Supreme Court in *San Antonio Independent School District* v. *Rodriguez*, 411 U.S. 1 (1973) where, in an interdistrict equity suit, the court held that "at least where wealth is involved, the Equal Protection Clause (of the Fourteenth Amendment) does not require the absolute equality or precisely equal advantages" (*San Antonio Independent School District* v. *Rodriguez*, 411 U.S. at 24).

Notwithstanding such important intervening developments, the two *Hobson* cases were surprisingly prescient on statistical, policy, and legal grounds, providing a rich source of analytical material for potential intra-district equity litigation. A few examples will suffice.

In his lengthy statistical analyses of spending patterns in the District of Columbia public schools, Judge Wright came to emphasize—perhaps overemphasize—the importance of salary differentials between rich white schools and poor black schools. Indeed, the bulk of *Hobson II* is directed at correcting this single inequity, which, according to one observer, left district officials free to increase other spending disparities while transferring teachers to meet the court's demand for equitable payrolls (Horowitz, 1977). Future litigants might wish to learn from this statistical tunnel vision and broaden the case at both the findings and remedies stages.

On policy grounds, Horowitz provides a fascinating glimpse of plaintiff Julius Hobson, Sr.'s intention for going into the suit—one that rings with surprising immediacy today—and Hobson's subsequent disappointment with the unintended consequences of his legal victory. Citing unpublished material, Horowitz discovered that Julius Hobson's real aim had been to equalize all expenditures "as a prelude to conferring on individual schools the power to spend their fixed, equal allocation as they saw fit" (Horowitz, 1977, p. 165; see also Baratz, 1975). Unfortunately, to implement the court's decision in *Hobson II*, more rather than less central control over school decision making became necessary—particularly but not exclusively authority over hiring decisions (see Horowitz, 1977, p. 165).

In terms of legal analysis, Judge Wright found rather astounding precedent in the Supreme Court's notorious decision in *Plessy* v. *Ferguson*, 163 U.S. 537 (1896), holding that racially "separate but equal" public facilities (originally streetcars or trolley cars) met constitutional standards (see *Hobson I*, 269 F. Supp. at 496). Writing more than 60 years later as a vocal supporter of the court's holding in *Brown* v. *Board of Education*, 347 U.S. 483 (1954), which reversed the abhorrent *Plessy* doctrine, Wright found in *Plessy* the seeds of what he called "this modern separate-but-equal rule" (*Hobson I* 269 F. Supp at 497). Recognizing that student transfers were not the solution to equalizing the predominantly black Washington, DC public schools, Wright sought fiscal equity rather than integration. A legal strategy that depends, even in part, on *Plessy* v. *Ferguson*, might be even more controversial today than when Judge Wright raised the point; but today's litigators should seriously consider the possibility, since it directly supports their position.

CONCLUSION

A growing consensus is emerging that schools are the appropriate level to assess the quality and equality of educational resources. As Thomas H. Jones explains: "The choice here (to analyze expenditures at the school level) not only has implications for data gathering and statistical analysis but it has profound implications for policy as well. For example, differences among the classrooms might imply that resources should be equalized not only to the school district level but even much closer to the actual delivery level for child services" (Jones, 1985, p. 203).

As *Hobson* v. *Hansen* illustrates, courts can also use school-site financial information to frame a legal decision and to present and monitor a possible remedy. The District of Columbia school board was forced, subsequent to the case, to assess staffing patterns in its schools since many of the high-cost senior staff worked in the middle-class (mainly white) schools, while less-experienced (and thus less-expensive) teachers were assigned to schools in the tougher, mainly African-American neighborhoods, creating part of the inequities discussed by Judge Wright.

Now that accurate school-site expenditure data are becoming readily available, we must wonder if the courts would be swayed by school-site financial information that showed the inequalities or inadequacies (or both) of resources actually reaching schools and students.

What if data showed that poor children were receiving an "inadequate" education because of the lack of resources at the district, school, and classroom levels? A pending New York state case, *The Campaign for Educational Equity* v. *The State of New York*, presumes that New York City public school students are being denied their constitutional rights because the City of New York contributes more tax money to the state than the state funds back for education. Conversely, what if some schools had adequate resources? Would a presumption of district wrongdoing or mismanagement adhere, undercutting claims of state-generated inequities?

We believe that such school-site inequity is common in hundreds, perhaps thousands, of school districts across America. But rather than advocate the proliferation of school-level equity litigation across the country, we would prefer to see state and district leadership (e.g., state commissioners, local superintendents, and school boards) assess their own patterns of spending and make improvements in school-site equity and efficiency without getting into the time-consuming, costly process of "going to court." Thus, another round of *Serrano*-style cases may not be necessary to increase school-site equity, using the latest models and methods.

Consistent with the current trend of pushing more and more educational decision making closer to the classroom, school-site spending patterns should stimulate informed debate over how school districts and individual schools actually allocate their resources. Concerns include: (a) the efficiency of central offices in managing costs and passing resources on to schools; (b) the ability of school-site leadership to manage their schools effectively and to fund students; (c) the effectiveness of teachers in gaining and using resources for quality classroom instruction; and (d) the determination of whether students receive the benefit of resources and are performing up to standard.

School-site expenditure data also facilitate closer analysis of where to spend money to get better results. Much of the debate about "does money matter?" is meaningless without better systems of data gathering and reporting information, such as analysis provided through In$ite. As data from Mayville and Westville illustrate, the first step was the building of a model that traced costs to children, with the school as the centerpiece. Next, the process led to equity and productivity function analysis (see Cooper & associates, 1994), indicating that more money in the classroom led to more learning and better test results. From here, we believe that the exploration of the relationship between resources and results should remain a top priority.

An expert in school-site decision making, Susan A. Mohrman, in her book, *School-Based Management: Organizing for High Performance* (1994), explains just how important this kind of analysis really is:

> Contributing to the success of the organization requires understanding its business elements, including its financial system, accounting practices, sources of revenue and expense, and the various constraints and requirements placed upon it by its different stakeholders. Again, this is an area where historically only managers have received training. Only if employees understand these issues are they able to make competent decisions that take the business realities and imperatives into account (p. 41).

We can thus link policy, information, and legal strategies as a means to force more resources to schools rather than district offices, where funding decisions are usually made—thus securing a better education for our children. Armed with tools and concepts discussed in this chapter, parents, teachers, and school administrators can find common cause for focusing and re-directing resources for the improved education of all students.

NOTES

1. *Serrano* v. *Priest,* 5 Cal.3d 584,487 P.2d 1241 (Super. Ct for Los Angeles County, CA, 1971).

2. Terms specific to In$ite, used in this section of the chapter, are printed with beginning capital letters to distinguish them from every day, generic usage.

REFERENCES

Baratz, J.C. (1975, January). Court decisions and educational change: A case history of the D.C. public schools, 1954-1974. *Journal of Law and Education, 4* , 63-80.

Berne, R. (1994, January 23). Study of the 32 community school districts in NYC public schools. *New York Times.* CY Section 13, pp. 12-14.

Berne, R., & Stiefel, L. (1984). *The measurement of equity in school finance.* Baltimore: The Johns Hopkins University Press.

Burrup, P., Brimley, E.V., Jr., & Garfield, R.R. (1996). *Financing education in a climate of change* (6th ed.). Needham Heights, MA: Simon & Schuster.

Coons, J.E., Clune, W.H., & Sugarman, S.D. (1970). *Private wealth and public education.* Cambridge, MA: Belknap Press of Harvard University.

Cooper, B.S. and Associates (1994, Summer). Making money matter in education: A microfinancial model for determining school-level allocations, efficiency and productivity. *Journal of Education Finance, 20* (1) , 66-87.

Hanushek, E.A. (1991). When school finance "reform" may not be a good policy. *Harvard Journal of Legislation, 28*, 423-456.

Hartman, W.T. (1981). Projecting special education costs. Paper 81-88. Stanford Institute for Research in Educational Finance and Governance, June 1981.

Hertert, L. (1995). Does equal funding for districts mean equal funding for classroom students? Evidence from California. In L.O. Picus & J.L. Wattenbarger (Eds.), *Where does the money go? Resource allocations in elementary and secondary schools* (pp. 71D84). 15th Annual Yearbook of the American Education Research Association. Thousand Oaks, CA: Corwin Press.

Horowitz, D.L. (1977). *The courts and social policy.* Washington, DC: The Brookings Institution.

Kozol, J. (1991). *Savage inequalities: Children in America's schools.* New York: Crown Publishers.

LaMorte, M.W. (1996). *School law: Cases and concepts* (5th ed.). Boston: Allyn & Bacon.

Mohrman, S.A. (1994). High involvement management in the private sector. In S.A.Mohrman, P. Wohlstetter, & Associates (Eds.), *School-based management: Organizing for high performance* (pp. 25D52). San Francisco: Jossey-Bass.

Mohrman, S.A., Wohlstetter, P., & Associates (Eds.) (1994). *School-based management: Organizing for high performance.* San Francisco: Jossey-Bass.

Odden, A.R. (1992). Toward the 21st century: A school-based finance. In A.R. Odden (Ed.), *Rethinking school finance: An agenda for the 1990s* (pp. 321-336). San Francisco: Jossey-Bass.

Odden, A.R., & Picus, L.O. (1992). *School finance: A policy perspective.* New York: Macmillan.

Sherman, J. (1996). *Assessing school-level equity in two states.* Reston, VA: Pelavin Research Institute of the American Institutes for Research (AIR).

Speakman, S.T., & Associates (1996). Bringing money to the classroom: A systemic resource allocations model applied to the New York City public schools. In L.O. Picus & J.L. Wattenbarger (Eds.), *Where does the money go? Resource allocations in elementary and secondary schools* (pp. 106-132). 15th Annual Yearbook of the American Education Finance Association. Thousand Oaks, CA: Corwin Press.

Wise, A.E. (1967). *Rich schools, poor schools: The promise of equal educational opportunity.* Chicago: University of Chicago Press.

PART III

IMPROVING SCHOOL PRODUCTIVITY

Chapter 8

WHERE'S THE MONEY GONE?

Richard Rothstein

When Benno Schmidt resigned Yale's presidency to head a private school network, he explained why he'd given up on public schools: "We have roughly doubled per-pupil spending (after inflation) in public schools since 1965," but the "nation's investment in educational improvement has produced very little return" (Schmidt, 1992).

This is a conventional claim of public school supporters and critics alike. School finance expert Allan Odden notes that "real" education expenditures increased by 58 percent in the 1960s, 27 percent in the 1970s, and 30 percent in the 1980s, "but student performance—and thus education productivity—have not improved that much" (Odden, 1992, pp. 10-11). According to a Brookings Institution report by John Chubb and Eric Hanushek, "since the Soviets launched... Sputnik,... real expenditures per student rose at an annual rate of three and three-fourths percent, nearly tripling between 1960 and 1988... Spending has nearly tripled and performance has dropped"

Advances in Educational Productivity, Volume 6, pages 197-216.

ISBN: 0-7623-0082-5

(Chubb & Hanushek, 1990). "Look at spending on public schools," *Public Interest* editor Irving Kristol concludes. "It goes up and up and up and the results go down and down and down" (DeParle, 1993).

The declining "productivity" claim is so well established that few analysts have sought empirical verification. Rather, the notion is but prelude to reform prescriptions; if, after all, everyone knows that growth in public education spending has outpaced any rise in school achievement, the challenge must be to design systems that use money more effectively, and there is no need to consider proposals for additional funds. Even many of those committed to public education assume that reform is urgently required to halt a scandalous waste.

The following pages summarize an Economic Policy Institute report that reexamines the apparent consensus that real school spending has roughly doubled in the last quarter century.[1] It concludes that while spending *has* risen substantially, the increase is both smaller and more complex than most assume: Real school spending increased by 61 percent from 1967 to 1991, only a little more than half the real growth conventionally assumed. And barely one-fourth of this increase was directed at "regular education," the traditional school activities whose outcomes can be measured in test scores, graduation rates, and so on.

While critics decry the large increase in school spending, newspaper headlines simultaneously proclaim stagnant student achievement. However, the prevailing view among education researchers seems to be that academic outcomes have mostly been stable over the full course of this period and minority students may have shown modest improvement, closing some of the gap with whites. If these claims of modest improvement are accurate, then it may be that regular education inputs and outcomes have each grown modestly. The following analysis' challenge to the myth of rapid growth in regular education spending does not entail a suggestion that school reforms are inadvisable. Instead, it argues that reforms are not likely to be well designed if they assume, based on exaggerated accounts of regular education expenditure growth, an unproven school productivity collapse.

Productivity measures are based on two components—what inputs are utilized, and what outcomes result. "Dollars spent per pupil" is often used to measure inputs. Measuring outputs is more complex since schools perform a variety of functions. Despite public fascination with test-score reports, education researchers and policy professionals lack agreement about appropriate definitions of outcomes, and about how well test scores or other indexes measure these outcomes. How should we measure children's health to ask whether school breakfast and lunch funds are well spent? What outcomes do we seek for severely mentally retarded or for trainable mentally handicapped children, to evaluate spending growth in special education? Do low-income students' graduation rates respond sufficiently to investments in compensatory education or dropout prevention programs? What academic progress do we expect for immigrant children whose families arrive with varied levels of literacy? How do we judge goals of vocational education?

No single measure, such as regular academic test scores, can accurately represent all school outcomes. To understand school spending and determine whether productivity is truly declining (i.e., whether additional dollars have been poorly used), we must link expenditures to the program and to the specific outcomes the spending was designed to enhance. If, for example, we discover that schools have used much of their new money to improve training of mentally handicapped youngsters, it would make no sense to judge the effectiveness of this spending by whether S.A.T. scores improved for the college bound.

The following analysis differs from conventional accounts of changes in school spending in two ways. First, it uses a more appropriate method to adjust for inflation to explain changes in spending over time. Second, it examines changes in school expenditure by program (e.g., regular education, special education, transportation, school lunches, etc.), not by function (instruction, support services, etc.) or by "object" (salaries, supplies, etc.) as is commonly done in education research and reporting.

The analysis categorizes expenditures by program for nine school districts in 1967 and 1991. The districts (identified by name in the accompanying tables) were chosen to mirror U.S. districts in size, expenditure growth rates, location, urbanicity, minority enrollment, and relative affluence.

Adjusting expenditures for inflation and categorizing them by program shows that:

- Real elementary and secondary per pupil spending grew by 73 percent in the nine districts between 1967 and 1991. Spending of all districts nationwide grew by 61 percent.
- The share of all spending for regular education (what most people think of as schools' normal academic function) declined from 80 percent in 1967, to 59 percent in 1991; nonetheless regular education per pupil expenditures grew 28 percent over the period; regular education received 26 percent of the net new money spent in 1991 compared to 1967 spending.
- Special education's share of all expenditures rose from 4 percent in 1967 to 17 percent in 1991.
- About 8 percent of net new money went to expansion of the school lunch and breakfast program. Another 7 percent went to attendance, dropout prevention, alternative instruction, and counseling.
- The growth of regular education spending was more marked in suburban than in urban districts. In urban districts studied, per pupil regular education spending grew hardly at all and might have declined in real terms if districts had not cut back on operations, maintenance, and general administration spending. But in some suburban districts, regular education resources grew substantially.

ADJUSTING EXPENDITURES FOR "INFLATION"

The claim that "real" per pupil spending doubled from 1967 to 1991 means that schools now utilize twice the inputs they utilized a quarter-century earlier. To say this, however, requires a calculation of what it would have cost, in 1991, to purchase exactly those inputs (teachers, textbooks, administrators, etc.) schools purchased in 1967. If the cost of 1967 resources (in "1991 dollars") was less than half of actual 1991 spending, we can say that per pupil expenditures did indeed double.

Such calculations reveal, however, that claims of "doubled" spending are inaccurate; real school spending actually grew by about 61 percent from 1967 to 1991, or about 40 percent less than most school critics claim. Here's why.

In 1967 American schools actually spent an average of $687 per pupil. By 1991 this had climbed to $5,566, a growth of 710 percent. Before critics make their "doubling" claim, they "adjust" nominal 1967 spending figures for inflation. But making this adjustment requires assumptions that education research has generally failed to examine.

Over any period, school expenditures may rise solely because of inflation, or they may rise because more resources—like teachers, textbooks, aides, and so on—are purchased; or from a combination of both factors. For example, if a district is faced with a 10 percent rise in the price of textbooks, it must increase textbook spending by 10 percent to provide students with the same number of textbooks. If college graduates in other fields get higher starting salaries, schools will have to increase salaries too, just to attract the same quality of teacher. The resulting pressure on school budgets is termed inflation—paying more for the same resources.

But if a district chooses to supply students with more (or better) textbooks, or to reduce class sizes by hiring more teachers, or to recruit higher quality teachers by paying salaries above the going "market" rate, these decisions represent "real" increases in school spending. To know how much real spending increased, we must separate increases that reflect decisions to purchase more resources from increases caused by inflation (price increases for the same resources).

This requires calculating what it would have cost, in 1990-1991, to purchase exactly those inputs (teachers, textbooks, administrators, etc.) schools purchased in 1966-1967. To make such calculations, it is commonplace to use the "consumer price index for all urban consumers." Using this "CPI-U", $687 in "1967 dollars" becomes $2,794 in "1991 dollars." In "real" terms, therefore, per pupil expenditures went from $2,794 to $5,566, or a quarter-century jump of 99 percent. As Benno Schmidt claimed, we "roughly doubled" real school spending.

But use of the CPI-U to adjust school spending overstates school spending growth. The inflation rate for school purchases is likely to be (and will continue to be) greater than average urban consumers' price inflation which the CPI-U is designed to measure.

The Bureau of Labor Statistics calculates the CPI-U by tracking actual price increases of all goods and services that "typical" urban consumers buy. They call this collection of goods and services a "market basket." But prices of some goods and services, in any period, go up faster than others. The CPI-U is only the "average" price increase of all items in the market basket, weighted by how much of that product or service the typical urban consumer buys.

Because schools are not "typical urban consumers," the "market basket" we use to calculate elementary and secondary education inflation should be different from the market basket we use to calculate consumer inflation. Schools, for example, rarely rent apartments or buy used cars, and urban consumers rarely hire teachers or buy textbooks.

Inflation rates vary widely among particular items. Medical inflation, for example, has been much greater than average consumer inflation, while prices of personal computers or VCR's have grown less than average consumer inflation—these prices have fallen, in fact. If, upon examination, we conclude that schools' market basket includes a greater proportion of services that behave like medical care and a smaller proportion of products that behave like VCRs than the proportions of these services and products included in the market basket of a typical urban consumer, then school inflation will be faster than the CPI-U.

The price of education services is, in fact, likely to rise faster than the price of an urban consumer's average purchases. Inflation in services typically exceeds inflation in commodities because productivity (the increase in output per employee-hour worked) is likely to grow more slowly in services. Productivity growth in manufacturing has allowed industrial firms to reduce their costs (or at least slow the growth in costs) and therefore increase the prices of manufactured products more slowly or not at all. In contrast, many service sector firms and institutions cannot easily substitute machinery for labor as manufacturers do; these organizations, for whom it's more difficult to achieve productivity growth, have had to increase prices faster than average. Oft-cited examples include barbers and orchestras: barbers can't greatly increase the number of haircuts they perform per hour and orchestras can't perform music with fewer musicians each year. These insights—that disparities in inflation mirror differences in productivity growth, and that industries (i.e., services, barbers, orchestras, and yes, schools) in which it's hard to achieve productivity growth will have higher-than-average inflation—are associated with the work of economist William Baumol (Baumol, 1967; Baumol, Blackman, & Wolff, 1989).

Education is subject to a "Baumol effect" because educational productivity improvements from cost reductions are difficult to achieve. In contrast, manufacturing industries can automate work and find efficiencies in use of materials; they thereby reduce resources needed in production. In the private sector the number of workers necessary to produce an average product fell roughly a third from 1967 to 1991. What would comparable labor productivity growth have looked like in schools? To oversimplify a bit, if the pupil-teacher ratio was 20:1 in 1967, and if schools experienced productivity growth the way the private sector did (by reducing labor inputs and using remaining inputs more efficiently), the pupil-teacher-ratio would have to rise to 26:1 in 1991 order to keep school costs from rising faster than average costs. If, on the other hand, we kept class sizes from increasing, per pupil spending would have to increase relative to consumer spending on manufactured products, and this increase would take place without any new real resources for schools!

Continuous industrial-like realization of cost efficiencies are difficult to imagine in schools, so it is inevitable that inflation in a low productivity-growth industry like education will be higher than inflation in an average industry experiencing average productivity gains. Real spending per pupil will rise even though more resources per pupil are not being provided. For this reason, use of the average inflation rate for consumer goods and services (the CPI-U) systematically understates the inflation facing school districts. It gives the impression that more of the nominal spending growth of 710 percent represents real new resources provided to districts for educating students, and less of the actual spending growth represents inflation, than was in fact the case.

Because of these considerations, the conclusions reported here do not assume an adjustment of 1967 school expenditure data with the CPI-U. Instead, the Economic Policy Institute, with assistance from the Bureau of Labor Statistics, developed a modified index (we call it the "net services" index, or NSI) to measure inflation in service industries like education. When we adjust schools' average per pupil spending of $687 in 1966-1967 by the NSI, we find that "real" spending in 1967 was $3,456. Real school spending growth is the amount of new money spent in excess of inflationary changes measured by the NSI— or 61 percent real growth.

ASSIGNING EXPENDITURES TO PROGRAMS

States and districts do not normally report school spending data by program type, although education researchers increasingly call on them to do so. So to understand which programs received which funds, we must first define programs and then categorize each expenditure of the nine sample districts by program and by employee type (teachers, aides, other professionals, other employees).

Others may distinguish programs differently, but the classification scheme summarized in Figure 8.1 is an attempt to most accurately reflect schools' varied outcome goals. Nonetheless, distortions are inevitably implied by any taxonomy of school programs. The goal can only be to minimize these distortions, not eliminate them.

BASIC EDUCATIONAL PROGRAMS

Regular Education
Vocational Education

PROGRAMS SERVING SPECIAL POPULATIONS

Special Education
Counseling and Dropout Prevention
- Alternative Education Instruction (for "At-Risk" Youth)
- Counseling, Attendance Control and Dropout Prevention

Bilingual Education
Compensatory Education

OTHER PROGRAMS

After-School Athletics
Regular Student Transportation
Regular Health and Psychological Programs
Food Services
Security and Violence Prevention

COSTS DISTRIBUTED PRO—RATA TO ABOVE PROGRAMS

Employee Benefits
Overhead
- General Administration
- School Administration
- Operations, Maintenance, Utilities

Figure 8.1. Major Expenditure Programs of Elementary and Secondary Education

For example, in a comparison like this one of expenditures in 1967 and 1991, before we can decide whether to assign expenditures to a special program or to regular education, we must first address whether the special program reflects a new or broadened school goal, or if changes have simply occurred in how similar students are categorized. In other words, do special student populations (e.g., students in special, bilingual, or compensatory education programs) mostly consist simply of children whom, in 1967, we considered "regular" students but who in 1991 have been reclassified, or do special student populations mostly represent a new challenge not confronted by American public schools in 1967?

The expectation that all elementary and secondary education spending should affect the outcomes (e.g., test scores) of regular education students rests on the former assumption. The implicit claim is that students in today's special education, bilingual, or compensatory education programs were, in 1967, mostly instructed as part of schools' regular program, and that the 1967 regular education outcomes from which we now expect improvements then included the outcomes of students similar to those who, in 1991, were in special programs. If this were the case, then any new resources given to these special programs represent only the reclassification of resources previously given to regular education. Changes in regular education outcomes, therefore, could be expected to measure the usefulness of these new resources.

This assumption may be true in some cases. There were some students in regular education programs in 1967 challenged by learning disabilities that today earn special education classifications; there were some students in regular education programs in 1967 who came from immigrant families not speaking English; and there were many students from poor minority families who struggled to graduate without compensatory assistance. If these descriptions truly characterize most students now in special programs, readers may fairly conclude that the analysis in these pages gives a false picture of changes in the programmatic divisions of education spending.

But on the whole, this is not generally the case. Most students served by special programs in 1991 would not have been in regular education in 1967, or if they were, they would not have remained in school for long. Some students enrolled in special programs in 1991 would not have been in school at all in 1967.

The claim that special programs represent simply a reclassification of regular students is most often made with respect to special education, and particularly with respect to children classified as "learning disabled." Yet while some learning-disabled children might be in regular classes were it not for the special education program, severely handicapped children would not be in schools at all, and would instead be cared for at home or in other institutions (as was the case in 1967).

Spending growth of special education masks complex changes in educational practice, medical technology, and social structure. In some cases, these changes reduced special education spending. Eradication of rubella reduced costs of special state schools for deaf children. By 1967 the number of children with severe orthopedic handicaps had been much reduced by polio vaccines, and polio was disappearing as a cause of special education enrollments.

In other cases, apparent growth of special education funds represents shifting public and private social resources, not net increases. Public schools now handle severely handicapped cases that previously were treated in noneducational public or private institutions. State schools for "palsied children" in 1967 were part of state hospital systems and would not normally be included in education expenditure data for that year. Public schools now accommodate students of chronological school age who are so developmentally retarded that they are not toilet-trained. In 1967 such children were often housed in private charitable institutions without benefit of school reimbursements. Children with less severe mental retardation were also cared for privately, and did not attend public schools.

In most cases, however, special education growth reflects real new spending, not shifts from other agencies. Partly this represents new sophistication in diagnosis of those who, in 1967, would have been deemed "slow" or unsuitable for academic education. Children with mild autism, severe dyslexia, or attention deficit disorders are examples.

New disabilities spur some expenditure growth. Children born to mothers on drugs may have higher incidence of learning disability. Children born prematurely with birth defects or low birth weights were, with modern medical technology, more likely in 1991 than in 1967 to survive to school age, but also more likely to need special education services.

There has also been growth in special education classifications of children with severe emotional and behavioral disorders. In part, these children may be products of family and community structures more dysfunctional than those from which 1967 children came. In part, growing special education expenditures may be due to a greater determination to prevent such children from dropping out of school.

Some special education cost increases result from opportunities to educate children who, without modern medical technology, would not have survived to school age in 1967. In some cases, these children survive with severe handicaps, but in others, there is simply a higher incidence of learning disability. Considering these children, it is apparent that even all learning-disabled classifications cannot be thought simply to be reclassifications of regular students, but instead are partly the net addition of a special population to school enrollment.

Nonetheless, some districts may have higher ratios of learning-disabled to severely handicapped children. For such districts, the categorization method described here may understate regular education spending growth. For others, with relatively fewer learning-disabled classifications but more severely handicapped or mentally retarded children, a decision to include in regular education the costs of educating the latter would inappropriately exaggerate the growth of regular education funds.

In deciding whether to create other special program categories to which 1991 expenditures should be assigned, this analysis makes similar difficult decisions about whether, on balance, students similar to those served would have been in regular education in 1967 or if such students were simply unlikely to have been part of elementary and secondary public institutions in 1967. In cases where the former seems to be more likely (for example, "gifted and talented" education), this analysis includes 1991 expenditures in regular education. In cases where the latter seems more likely, expenditures are distinguished as part of a special program.

Thus, there were more first-generation immigrant children in 1991 schools than there were in 1967, and after the abolition of "national origin" immigration quotas in 1965, a greater proportion of immigrant children came from semiliterate families in 1991 than in 1967. The rise of undocumented immigration from Mexico and the Caribbean, the surge of Indochinese refugees in the 1970s and 1980s, and recent growth of undocumented Asian immigration accentuate this trend. Immigrant children, in the absence of bilingual programs, might sit in regular classes unable to benefit, so it would be inaccurate to consider expenditures these children require as replacing regular education. A contrary decision—to include bilingual education program spending as part of regular education in 1991—would imply that in 1967 schools educated similar children as part of regular education programs and without special help. Certainly, some immigrant children were successful in schools' regular programs in 1967. But the larger share of 1991's non-English speaking schoolchildren had few counterparts in 1967 schools. The resources devoted to these students in 1991, and their outcomes, should therefore be accounted for separately.

In the case of some other special populations ("at risk" or compensatory education students, or students who benefit from desegregation expenditures), some students may have been in regular classes if not for special programs, but other students might have dropped out. Thus, this analysis includes as part of a special program for "at risk" youth actual instructional expenditures for teachers and aides in alternative schools for students who would otherwise drop out, evening schools for secondary students who work or care for siblings during regular school hours, and disciplinary schools for students who voluntarily or involuntarily separate from regular schools. These instructional expenditures are not included in regular education because costs of alternative secondary schools would mostly not revert to regular education were these schools not in existence. Instead, these students would likely drop out of school, unserved by any educational program.

Dropout rates were probably much higher in 1967 for minority and low-income children whom these programs now target (although dropout data are the most unreliable of education statistics). For example, many at-risk students now live in communities which had few such students before the great northern migration of rural Southern blacks, a migration still in process during the early years of the study on which this analysis is based. Many of these students in 1967 attended small rural schools in small rural districts not comparable to the urban and suburban districts where they attend school today. Others, if they attended large districts, were often segregated and did not receive the same academic resources as white students. In Anne Arundel, Maryland, one of the districts on which this analysis is based, in 1967 college degrees were not required of teachers in segregated schools. This analysis also includes data for East Baton Rouge, Louisiana schools which in 1967 are the compilation of separate data for "Negro Schools" and "White Schools." Per pupil spending was much higher in the latter. Thus, to say that in the absence of compensatory or desegregation programs these students "would have" been in regular education, and that regular student average outcomes should improve from 1967 to 1991 as a result of spending on minority and impoverished students, is a farther stretch than the facts justify. On the other hand, it must be acknowledged that this taxonomy creates a less serious distortion in the opposite direction.

There is no perfect way to resolve the conflict. In a district where magnet schools were established pursuant to a desegregation lawsuit, this categorization scheme considers as an instructional expense of desegregation the additional, but not the base expenditures of these schools. Because the court ordered smaller classes in magnet than in regular schools (to make magnets more attractive to an integrated student body), additional costs of lower class sizes are categorized an expense of desegregation, not of regular education. (Basic instructional costs are considered regular education expenditures.) If these monies now spent in magnets would, in the absence of a court order, otherwise be spent on the same students but in regular educational programs, the methodology of this analysis overstates the extent to which desegregation programs compete with regular education for funding. If, on the other hand, in the absence of magnet schools white students would leave the affected district (either for private schools or suburban communities), and if minority students in racially isolated schools would then receive inferior educational services, the methodology used here is more appropriate.

No classification rules can accommodate the many circumstances involved. This analysis attempts to categorize spending in ways that do not understate regular education's share of school spending in 1991. Some overstatement is likely since many district finance records do not identify special program costs unless federal or state eligibility exists for reimbursement. Therefore, this report probably categorizes some expenditures as regular education when better knowledge of their uses might have suggested a different program.

The methodology used here also overstates the share of resources for regular education because it cannot account for the time of regular teachers devoted to special programs. When special education students are mainstreamed in regular classes for part of the school day, some portion of teachers' time for regular academic tasks is redirected. All compensation of these teachers, however, is still recorded as a regular education expense. While the research team that assembled data for this report conducted interviews to understand teachers' time allocations, the interview data was too subjective to form a basis for decisions about spending allocation. But regular education teachers often insist that they spend more time today on special programs and less on regular education than they did 25 years ago.

Those who disagree with this methodology may still find the data reported here useful. For example, readers who believe that "compensatory education" should not be distinguished from regular education can sum the rows for these programs in Table 8.1. They will find that regular education (compensatory education included) had an 85 percent share of all spending in 1967 but only a 62.1 percent share in 1991. In this case, regular education's drop in share was more precipitous than the table suggests (because federal Chapter 1 funds were a smaller share of district spending in 1991 than in 1967). With other combinations of data from the table, on the other hand, regular education's drop in share could seem slightly less precipitous, but the broad picture would remain.

ASSIGNING ADMINISTRATIVE EXPENSES TO PROGRAMS

Here is one example of how this particular scheme approaches policy issues implicit in a decision about classification: Former Education Secretary William Bennett's accusation that schools waste funds in an "administrative blob" has led many researchers to focus on distinguishing administrative from classroom costs. They generally find administrative costs (central and school level) of 8 to 17 percent. The analysis reported here, on the other hand, assigns many expenditures, often termed "administrative," to programs.

Implicit in the conventional approach is a model that likens classrooms to factory floors where "direct" teaching labor carries out production, other functions providing indirect support. But, as in manufacturing, schools do not succeed only as "direct to indirect" ratios grow. Success also depends on the intelligence with which the enterprise is planned and coordinated, as well as on the product mix created. The implicit notion in much contemporary education debate that classrooms are "profit centers," while curriculum libraries or school buses are "cost centers," prevents thoughtful analysis of programmatic productivity. Central office development of curriculum guides, for example, and teachers' transmission of this curriculum, are equally necessary to instruction. Either may be conducted effectively or wastefully. By calculating total costs for each program, including its administrative costs, the analysis reported here does not suggest that leadership funds are well spent, any more than identification of classroom expenditures suggests the most effective teaching techniques. This cannot be determined by finance analyses alone and must be addressed in separate inquiries.

The approach used here creates categories not comparable to those of studies specifically segregating administrative costs because this analysis distinguishes administrators readily identifiable with a particular program from those charged with overall direction of the school enterprise. This report assigns the former exclusively to the program with which they are identified; only the latter are included in a general administrative "overhead" or indirect category, ultimately allocated to programs in proportion to each program's "direct" expenditures.

For example, this analysis considers salaries of student transportation directors (and of their secretaries), or of programmers planning bus routes, part of a student transportation program. Conventional analysis treats this staff as general "administration," while bus drivers' salaries are transportation expenditures. This report, on the other hand, considers directors of special education as part of the special education program and district curriculum writers as part of the regular education program.

Because this analysis calls "general administration" only those administrators with overall responsibility (like the superintendent or finance and personnel staff), "administration" is a smaller part of the totals reported here than the "central administration" category with which education analysts are familiar. The focus in this analysis on program asks different questions.

School officials claim that general and site administrators spend disproportionate time on special populations and problems. If so, allocating indirect administrative costs in proportion to identifiable program expenditures

Table 8.1. Shares of Total Spending for Each Program, Average of Nine Districts, 1967 and 1991 (Programs Listed in Order of 1991 Share of Total Spending)

	Share of Total Per Pupil Spending		*Change In Share,*
Program	*1967*	*1991*	*1967-1991*
Regular Education	79.6%	58.8%	-20.9
Special Education	3.7	17.0	13.3
Compensatory Education	5.4	4.3	-1.1
Attendance, Counseling, Dropout Prevention, Alternative Education	2.1	4.1	2.0
Food Services	2.0	4.1	2.1
Regular Student Transportation	3.9	3.4	-0.5
Vocational Education	1.4	3.0	1.6
Bilingual Education	0.3	1.8	1.5
Desegregation	0.0	1.6	1.6
Regular Health and Psychological Services	1.3	0.9	-0.3
After-School Athletics	0.4	0.7	0.3
Security and Violence Prevention	0.1	0.4	0.3
All Programs	100.0	100.0	
Overhead Allocated to Above Programs:			
General and School Administration	9.4	9.7	0.2
Operations and Maintenance	15.7	14.3	-1.4

overstates rising costs of regular education and understates growth of special populations' costs. In preparing this analysis, the EPI research team did interview administrators to understand their time allocations, but concluded that we could not make adjustments for this without unacceptable guesswork.

Categorizing expenditures by program creates results that will be unfamiliar. First, even those who specialize in understanding school finance will be unaccustomed to seeing programmatic expenditures, rather than reports of "function" or "object." Second, because this analysis includes pro-rata shares of state government expenditures (like textbook selection, testing, special schools, teacher certification, retirement fund contributions), total expenditure figures obtained here differ even from districts' own reports. Because the thousands of separate category decisions made for this analysis are not easily comparable to districts' own reports, the data summarized here must be subject to verification by other researchers. Therefore, the districts in the study sample were not offered anonymity, and the Economic Policy Institute will make arrangements for qualified researchers who wish to review these data to test the validity of the conclusions reported here.

CHANGES IN PROGRAM SHARES OF EDUCATION SPENDING

Table 8.1 describes, for the nine districts, changes in shares of total spending for each program. The table shows that special education's share increased the most, from less than 4 percent of all spending in 1967 to 17 percent in 1991. In contrast, the share of funds going to regular education declined. In 1967 regular education consumed almost all elementary and secondary education dollars—80 percent went to regular classroom teaching, school libraries, textbooks, curriculum development and teacher training, and the associated share of maintenance and administrative costs. Regular education saw its share of elementary and secondary spending fall to 59 percent in 1991. Table 8.2 describes the distribution of net new money in 1991 in the nine districts. Special education took the largest share of net new money: 38 percent. Regular education received only 26 percent of net new money.

Tables 8.3 and 8.4 compare changing shares of total spending and the amounts of net new money going to regular and special education, the two largest spending categories for each district. They show that each district in the sample decreased its share of money spent on regular education, and increased the share spent on special education.

Table 8.2. Shares of Net New Spending by Program, 1967-1991 (Programs Listed in Order of Share of Net New Per Pupil Spending)

Program	*Share of Net New Per Pupil Spending, 1991 Nine-District Average*
Special Education	38.0%
Regular Education	25.9
Food Services	7.5
Attendance, Counseling, Dropout Prevention, and Alternative Education	7.4
Vocational Education	5.2
Desegregation	4.1
Bilingual Education	3.9
Compensatory Education	2.9
Regular Student Transportation	2.8
After-School Athletics	1.1
Security and Violence Prevention	0.9
Regular Health and Psychological Services	0.5
All Programs	100.0
Overhead Allocated to Above Programs:	
General and School Administration	9.8
Operations and Maintenance	12.3

Table 8.3. Changes in Regular Education Share of Per Pupil Spending, 1967-1991 (Listed in Order of 1991 Share for Regular Education)

	Regular Education Share of Per Pupil Spending			
District	*1967*	*1991*	Change *1967-1991*	Regular Education Share of Net New Per Pupil *Spending, 1967-1991*
Bettendorf, IA	92%	72%	-20	36%
Boulder, CO	84	64	-20	9
Anne Arundel, MD	83	62	-21	41
Spring Branch. TX	83	60	-23	44
Middletown, NY	78	59	-19	35
East Baton Rouge, LA	76	57	-18	22
Claiborne, TN	57	53	-4	46
Fall River, MA	76	51	-25	4
Los Angeles, CA	87	51	-36	-5
Average	80	59	-21	26

Table 8.4. Changes in Special Education Share of Per Pupil Spending, 1967-1991 (Listed in Order of 1991 share for Special Education)

	Special Education Share of Per Pupil Spending			
District	*1967*	*1991*	*Change 1967-1991*	*Share of Net New Per Pupil Spending, 1967-1991*
Fall River, MA	8%	22%	14	49%
Middletown, NY	2	22	20	46
Los Angeles, CA	2	18	16	42
Anne Arundel, MD	3	18	15	33
Boulder, CO	4	16	12	47
East Baton Rouge, LA	4	16	12	40
Bettendorf, IA	3	16	12	39
Spring Branch, TX	3	13	10	19
Claiborne, TN	4	12	9	26
Average	4	17	13	38

REGULAR EDUCATION SPENDING

The decline in the share of total spending for regular education does not mean that real per pupil spending for regular education fell. A smaller share of a larger whole can still provide increases. Because total per pupil spending grew by 73 percent, districts could increase their real regular education per pupil spending despite regular education's reduced relative priority.

Table 8.5 shows the real per pupil growth of regular education spending in the nine districts. The range is broad, from a high of 77.9 percent in Spring Branch to a decline of 3.5 percent in Los Angeles. The average was a 28 percent increase. Table 8.6 distinguishes direct from indirect (general administration, operation, and maintenance) expenses in regular education. It reveals that the 3.5 percent decline in Los Angeles's per pupil spending on regular education (an average annual decline of 0.1%) partly reflects a much larger decline (an annual average of 2.7%) in indirect operations and maintenance expenditures. However, even without this indirect overhead allocation, the direct per pupil expenditures for regular education in Los Angeles increased only by 9 percent from 1967 to 1991, an average annual rate of only 0.3 percent. This is significantly less than the average for the nine districts (where direct per pupil expenditures for regular education, exclusive of administrative, operations and maintenance overhead, grew by an annual average of 1.1%).

Table 8.5. Growth in Real Regular Education Per Pupil Spending

	Regular Education Spending Per Pupil ($1991)		*Total Percent Change*	*Average Annual Percent Change*
	1967	*1991*		
Spring Branch	1,825	3,247	77.9%	2.4%
Anne Arundel	2,513	3,780	50.4	1.7
Claiborne	1,027	1,524	48.4	1.7
Middletown	3,424	4,691	37.0	1.3
Bettendorf	2,674	3,229	20.8	0.8
East Baton Rouge	2,096	2,424	15.7	0.6
Boulder	3,189	3,317	4.0	0.2
Fall River	2,279	2,345	2.9	0.1
Los Angeles	3,118	3,010	-3.5	-0.1
Average Change			28.2	1.0

If this is typical of other urban megadistricts, it might help explain why there is such great concern about academic outcomes of these districts. Other urban districts in the sample, Fall River and East Baton Rouge, also saw real direct per pupil regular education costs grow by less than the sample average, and this was also partly a result of drastic reductions in maintenance expenses (and in Fall River administrative expenditures as well) attributed to regular education.

In contrast, suburban districts like Spring Branch, Anne Arundel, and Middletown saw direct regular education per pupil spending grow at average annual rates of 2.4, 1.7, and 1.4 percent, respectively.

Table 8.6. Growth in Real Regular Education Per Pupil Spending, 1967-1991, by Type of Spending (Average Annual Percent Change)

		Indirect		
	Direct	*General Administration*	*Operations and Maintenance*	*Total*
Spring Branch	2.4%	3.6%	2.1%	2.4%
Anne Arundel	1.7	1.8	1.8	1.7
Claiborne	1.4	4.7	2.4	1.7
Middletown	1.4	0.7	1.2	1.3
Bettendorf	0.9	0.8	0.3	0.8
East Baton Rouge	0.8	0.8	-0.4	0.6
Boulder	0.0	0.2	0.7	0.2
Fall River	0.4	-1.2	-0.5	0.1
Los Angeles	0.3	-0.4	-2.7	-0.1
Average	1.1	1.6	0.8	1.0

The 28 percent average real growth (for the nine districts) of regular education spending was consistent with dollar allocations between teaching, administration, instructional materials, and other regular education functions changing little from 1967 to 1991. As Table 8.7 shows, instructional staff (teachers and aides) consumed about two-thirds of regular education spending in both 1967 and 1991, though there was a slight shift from teachers to aides (the magnitude of this shift varying greatly by district). Despite availability of instructional technology, it appears that districts did not invest a greater share of their budgets in instructional materials and equipment. Staff development and curriculum planning grew from 2 to 3 percent of regular education budgets.

SPECIAL EDUCATION SPENDING

The biggest share of new K-12 money has gone for special education. In 1967 the nine sample districts spent less than 4 percent of total resources on special education. In 1991 they spent 17 percent. Special education growth consumed 38 percent of net new funds in 1991.

As discussed earlier, growth in special education spending may in some cases (but probably not too often) mask shifting regular education funds, a willingness by some districts to classify as "special" children whom in 1967 educators considered within the normal span of learning styles teachers confront. New classifications may result from greater sensitivity to causes of learning difficulties, desire to relieve regular teachers of disciplinary problems or the burden of special attention some children require, or simply because state reimbursement formulae allot more dollars per pupil for special education students. Lines between special and regular education placements are indistinct; wide variation in district practices are evident. It is widely believed that northeastern districts are more prone to claim that children with learning problems need "special education." The data collected for this report may support this: as displayed in Table 8.4, in 1991 Fall River and Middletown each spent 22 percent of all funds on special education, more than other sample districts. Claiborne and Spring Branch spent 12 and 13 percent, respectively; remaining districts spent 16 to 18 percent.

It will be difficult for education researchers to evaluate whether this substantial new investment in special education has been an effective means to improve outcomes, because outcomes in special education are especially difficult to measure. "Special education" really comprises many programs, each with different outcome goals. Since each special student must have an individualized education plan that specifies outcomes sought and resources and strategies to achieve them, outcome targets vary greatly by the nature of a child's disability.

Table 8.7. Changes in Share of Regular Education Per Pupil Spending by Category

	Share of Total Regular Education Per Pupil Spending 1967	1991	*Change*	*Real Growth in Per Pupil Spending,* 1967-1991 (% Change)*
Teachers (Including Substitutes)	67.5%	65.1%	-2.3	23.3
Aides	0.1	1.6	1.6	
Summer School	0.3	0.5	0.2	
Tuition and Contracted Services	0.1	0.2	0.2	
Staff Development and Curriculum Planning	1.9	3.2	1.2	
Library	1.4	2.8	1.4	
Textbooks and Instructional Equipment	3.6	2.6	-1.0	11.2
Overhead (General Administration Operations, Maintenance)	25.1	24.0	-1.2	27.9
Total	100.0	100.0		28.2

Note: * Real changes are not meaningful where 1967 amounts in some districts are zero or close to zero.

FURTHER RESEARCH

This report takes very preliminary steps in research areas previously ignored by the education research community. Elaboration and refinement of the "net services index" could provide more realistic pictures of education spending changes in other districts and time periods, as well as permit more realistic comparisons of resource growth in different districts and regions. The classification of expenditures by program for other districts and other time periods could help confirm whether the initial conclusions of the EPI report are valid. A uniform programmatic accounting system, if adopted by districts, would help policymakers learn where education money is spent. And development of outcome measures for schools' many programs is needed before programmatic expenditure data can be used in discussions of school productivity.

NOTE

1. This chapter is based on a 1995 Economic Policy Institute report, by Rothstein with Miles (1995). The full report can be obtained from the Economic Policy Institute at 800-374-4844.

REFERENCES

Baumol, W. (1967, June). Macroeconomics of unbalanced growth: The anatomy of urban crisis. *American Economic Review, 57*, 415-26.

Baumol, W., Blackman, S.B., & Wolff, E,N. (1989). *Productivity and American leadership.* Cambridge, MA: MIT Press.

Chubb, J.E., & Hanushek, E.A. (1990). Reforming educational reform. In H.J. Aaron (Ed.), *Setting national priorities. Policy for the nineties.* Washington, DC: The Brookings Institution.

DeParle, J. (1993, March 2). Social investment programs: Comparing the past with the promised payoff. *The New York Times.*

Odden, A.R. (Ed.) (1992). *Rethinking school finance. An agenda for the 1990s.* San Francisco: Jossey-Bass Publishers.

Rothstein, R., with Miles, K.H. (1995). *Where's the money gone? Changes in the level and composition of education spending, 1967-1991.* Washington, DC: Economic Policy Institute.

Schmidt, B.C., Jr. (1992, June 5). Educational innovation for profit. *The Wall Street Journal.*

Chapter 9

ARE PRODUCTIVITY AND EQUITY COMPATIBLE?

A CASE STUDY OF TEXAS SCHOOL FINANCE

Lawrence O. Picus

> In Senate Bill 7, the Legislature equates the provision of a "general diffusion of knowledge" with the provision of an accredited education. The accountability regime set forth in Chapter 35, we conclude, meets the Legislature's constitutional obligation to provide suitably for a general diffusion of knowledge statewide.
>
> *Edgewood I* and *Edgewood II* also require financial efficiency; that is, districts must have substantially equal access to funding up to the legislatively defined level that achieves the constitutional mandate of a general diffusion of knowledge. Unlike the school finance systems at issue in *Edgewood I* and *Edgewood II*, we conclude that the system established by Senate Bill 7 is financially efficient (Texas Supreme Court in *Edgewood IV*).[1]

With these words, the State Supreme Court of Texas ended over a decade of litigation and uncertainty over school funding, leaving legislators, educators and taxpayers to sort out the implications of Senate Bill 7, the school finance

Advances in Educational Productivity, Volume 6, pages 217-242.

ISBN: 0-7623-0082-5

distribution formula enacted a year and a half earlier, and now certain to remain the law guiding school funding issues for the foreseeable future. Since it was first filed in 1984, the *Edgewood* suit in its various forms has caused the state to make three major revisions to its school finance system. Along the way some school districts have been forced to make dramatic program cuts while other districts have seen their per pupil revenues increase by as much as $1,000 per pupil from one year to the next.[2]

Throughout this struggle the goal of school finance reformers has been to improve the equity of funding for the 1,044 school districts across the state of Texas. As established by the Texas Supreme Court in its first *Edgewood* ruling, equity implied elimination of the correlation between the per pupil wealth or property value of school districts and the per pupil revenues of the districts.[3] A number of earlier studies have shown improvements over time in the standard measures of school finance equity in Texas, and in fact, Picus and Hertert (1993a) and Picus and Toenjes (1994) argued that when compared to the equity of other states, Texas compared quite favorably.[4] The question in Texas was, what standard of equity did the state constitution require? In the end, the high court appears to have backed down from earlier *Edgewood* rulings in which they stated that achieving equity required eliminating the relationship between revenue and wealth was necessary for 100 percent of the districts in the state. In their last ruling the court concluded that the distribution of funds resulting from Senate Bill 7 meets the constitutional requirement.

The court's decision ended the *Edgewood* litigation. Future complaints about the inequitable distribution of funds in Texas must start over in the courts. As a result, this is an ideal time to take one more look at the distribution of funds to school districts across the state, and consider the equity of the formula established under Senate Bill 7. This paper summarizes earlier studies on Texas school finance equity, considers the equity of the current formula under a number of different alternatives, and projects the equity of future scenarios through the use of a funding simulation. The first part of the paper offers a history of Texas school finance over the last 25 years. This is followed by an analysis of the equity of the different approaches to school finance that have been tried in the state. The third section extends the analysis of equity for the 1992-1993 and 1993-1994 school years, considering the implications of alternative adjustments for differences in the cost of education across districts, adjustments for school district size, and alternative ways to count pupils with special needs. The fourth section presents the results of some simulations estimating school finance equity when Senate Bill 7 is fully implemented in 1996-1997. Finally, the last section of the paper presents some conclusions and final thoughts.

TEXAS SCHOOL FINANCE: A BRIEF HISTORY

In broad terms, the Texas school finance formula can be viewed as a two-tier finance system. The first tier is a typical foundation program with a basic allotment driven by a pupil count. The second tier is a guaranteed yield program offering districts a guaranteed revenue per penny of tax effort. In the past, districts have been allowed to levy taxes beyond the tax rate guaranteed by the state in the second tier. This yield was not equalized. Although the details of the school finance formula have changed in recent years, these basic levels have remained part of the system. Below, modifications to the Texas school funding formula are described within the context of the major court decisions that have taken place over the years.

Rodriguez v. *San Antonio*

Rodriguez v. *San Antonio* was the first of two major challenges to the Texas school finance system in the past 20 years. In 1971 a federal district court found that the state's school finance system included wide variation in educational expenditures directly attributable to local property wealth per pupil, and was in violation of the equal protection clause of the United States Constitution.[5] The case was appealed directly to the United States Supreme Court, which ruled in 1973 that the Texas system was constitutional in a narrow five to four decision.[6] Although this legal challenge was unsuccessful in requiring changes in the way Texas funded public education, by the end of 1984 a number of substantial changes in the school finance structure had been implemented, many of them aimed at improving the equity of the system.

Edgewood I

In 1984 a group of poor school districts, including the Edgewood Independent School District, filed the second major law suit, this time in state court, alleging that interdistrict expenditure differences violated their rights under the Texas constitution.[7] The plaintiffs agreed to hold their action pending the outcome of a special session of the Texas Legislature. In July 1984 the legislature enacted House Bill 72 which included a number of significant improvements in the equity of the school funding system. Verstegen found that the new system substantially improved Texas school finance equity, stating "the data indicated improvement over time in the long struggle of equalizing resources in the State of Texas and a robust response by lawmakers to issues raised by *Rodriguez*" (1987, p. 330).

Despite these improvements, the plaintiffs refiled the suit in 1985 as *Edgewood* v. *Kirby*.[8] In June 1987 a district court ruled the state's public school finance system unconstitutional. The decision was reversed by an appellate court in October 1989, but the Texas Supreme Court later unanimously upheld the district court, finding that the school finance system violated the state constitution's "efficiency" provision.

Edgewood II

The legislative response to this ruling was Senate Bill 1, enacted in early June 1990. The system created by Senate Bill 1 was not developed easily. The legislature and the governor fought with each other over the level of taxes to be raised, and other accountability and efficiency issues (Hobby & Walker, 1991). Senate Bill 1 was found unconstitutional by a state district court in July 1990. This decision was upheld in early 1991 by the Texas Supreme Court which ruled that Senate Bill 1 "leaves essentially intact the same funding system with the same deficiencies we received in Edgewood I."[9]

Following a clarifying opinion, known as *Edgewood IIa*, the legislature passed Senate Bill 351 on April 22, 1991. This legislation continued the three-tier funding formula used in Senate Bill 1, but the central feature of the first tier was the use of County Education Districts (CEDs) to raise local property taxes. These districts were designed to get around the prohibition on statewide recapture in the Texas Constitution by levying and distributing a portion of the school property tax within counties or groups of counties. As Toenjes (1991) points out, this countywide equalization was mathematically identical to a state recapture program, but was politically more acceptable and designed to get around a 1932 court ruling that held recapture to be unconstitutional.[10]

The second tier of the program was a guaranteed yield program that provided equalization for up to $0.45 of local levy effort beyond the CED property tax. For 1991-1992 the state guaranteed that each district would be able to raise $21.50 per pupil for each penny of tax effort, up to a maximum of $0.45. This guarantee was slated to increase to $28.00 per pupil for each penny of effort by 1994-1995.

The third tier was an unequalized local property tax. School property taxes were capped at $1.50. In 1991-1992 the CED tax rate was $0.72. With the next $0.45 equalized through tier two, the maximum unequalized school property tax rate was $0.33. Senate Bill 351 was to be fully implemented in 1994-1995. At that time, the CED tax rate would have been $1.00, and with the $0.45 equalized yield in the second tier, the unequalized portion of the school tax rate would only have been $0.05.

Edgewood III

Like its predecessors, Senate Bill 351 was challenged in the courts.[11] Although a number of issues were brought forward, the court initially focused only on the constitutionality of the CED tax collection system, with all other questions reserved for separate consideration. In August 1991 the district court ruled the CEDs constitutional. On direct appeal the Texas Supreme Court on January 30, 1992 held the CED tax system unconstitutional. The court's ruling came one day before property taxes were due. In a surprising decision the court deferred the effect of its ruling until June 1, 1993, in effect sanctioning the CED tax for both fiscal years 1991-1992 (the current year) and for 1992-1993, the next year. This gave the legislature until June 1993 to find an alternative.

There was talk of a special legislative session in May of 1992, but it never took place. A special session to consider school finance options was called immediately following the November 1992 general election, but during a three-week period, no agreement was reached. The matter became the responsibility of the new legislature which began meeting in January 1993. Although there was new leadership in the House, and a change in the chair of the Senate education committee, a solution to the school finance problem did not readily materialize. Eventually, three constitutional amendments were placed before the Texas voters. The first amendment proposed making the existing CED tax system constitutional, and proposed a statewide recapture of up to 2.75 percent of total state and local expenditures for education. This would have transferred approximately $400 million from the wealthiest districts to the poorer districts in the state and made recapture legal in Texas. The second amendment would have relaxed requirements on local districts resulting from unfunded state mandates, and the third dealt with borrowing for school facilities aid. The voters overwhelmingly rejected all three referendums on May 1, 1993, leaving the legislature with only one month to develop a solution. The result was Senate Bill 7.

Senate Bill 7 is a complex bill containing a number of important provisions affecting the finance and operation of Texas school districts. However, the most important part of this legislation has to do with its method to reduce wealth variations across school districts. Each of the 98 districts whose property wealth exceeded $280,000 per pupil was required to reduce its wealth to no more than $280,000.[12] There were five voluntary options available to these districts. According to Walker (1993), they are as follows:

1. *Consolidation by Agreement:* The boards of trustees of two or more districts may consolidate by agreement provided the property wealth of the consolidated district is no more than the equalized wealth level, defined in Senate Bill 7 as $280,000 per student in weighted average daily attendance (WADA). The districts do not have to be contiguous, and an election is not required.
2. *Detachment and Annexation by Agreement*: The boards of trustees of two districts may agree to the transfer of property from one district to another district for the purpose of taxation. The district from which the property is transferred must reduce its property wealth below $280,000

per WADA, and the receiving district's wealth can be no more than that guaranteed under the guaranteed yield program, or $205,500 in 1993-1994. Voter approval is not required for this option, but the property transfer is permanent.

3. *Purchase of Attendance Credits:* Under this option a district may purchase attendance credits from the state (i.e., send money to the state), effectively increasing its WADA until its property value is reduced to $280,000 per WADA. The cost of attendance credits are the greater of (a) a district's tax revenue per WADA, or (b) the state average tax revenue per WADA for the preceding school year. This option effectively amounts to state recapture, and local voter approval is required before a district can purchase the attendance credits. Interestingly, the recaptured funds include a portion of those amounts levied for debt service purposes.
4. *Contract for Education of Nonresident Students*: Under this option a district may pay to educate children in another district such that the WADA it is paying for, added to its own WADA, results in a total property wealth no greater than $280,000 per WADA. This is similar to option 3, only the funds are given to another district rather than to the state. The recapture effect is identical since the required expenditure is the state average total tax revenue per WADA for the preceding school year unless the commissioner certifies that a quality education can be delivered for a lesser amount in the receiving district. This option also requires approval of the voters of the purchasing district.
5. *Tax Base Consolidation:* Under this option two or more districts would agree to consolidate to create a taxing district similar to the CEDs. Voter approval must be received from all participating districts during the same election, and the entire tax rate would be levied by the consolidated district, unlike the CEDs which only levied $1.00 of the $1.50 tax levy.

If districts did not reach agreement on how to reduce their property value by September 1, 1993, then the commissioner was empowered to do so for the districts. The commissioner's options were limited to transferring property from wealthy districts to poor districts using a set of guidelines established in the legislation. Clark (1993) shows that all 98 districts were able to find a way to reduce their wealth below the $280,000 per pupil threshold. Among the 98 districts, 52 elected to purchase attendance credits from the state, eight chose to pay for education of students in other districts, 37 voted favorably for both options, and one district voted for voluntary detachment and annexation of property. In all cases, the wealthy districts were able to convince overwhelming majorities of voters to approve these options. It is speculated that the reason for this success was because districts felt that if they gave money to the state or to other districts, and Senate Bill 7 were declared unconstitutional, they would be able to keep the funds in future years, but if the property were annexed to another district, it might not be possible to get it back in the future.

Another issue that Senate Bill 7 attempted to resolve was proration. Proration is the term used in Texas to reduce school funding allocations when the state does not appropriate a sufficient amount to fully fund its commitment. In 1991-1992 proration amounted to approximately $300 million, and in 1992-1993 it was over $600 million. Under Senate Bill 351 a district's proration was directly related to its property wealth. Consequently, wealthier districts received larger per pupil reductions in funding than did poor districts. Picus and Hertert (1993b) found that the impact of this proration formula on the equity of the system focused on the very wealthiest districts in the state. Because the 109 wealthiest districts did not receive any state equalization aid, their state funds could not be reduced. Moreover, there were another 121 districts who received less equalization aid than their share of the proration. Since the state cannot take away money it has not given the districts, the proration that would have been assigned to these districts was apportioned among the remaining districts.

Picus and Hertert (1993b) found that, for the most part, this scheme of proration had a limited impact on measures of equity. With the exception of the Gini coefficient, all of the equity statistics calculated by Picus and Hertert were weighted by the number of pupils in the state. The Gini coefficient calculated by Picus and Hertert was a district-level variable. Because spending was not reduced in 109 districts, and only reduced marginally in 121 more, and because the wealthiest districts that were not part of the aforementioned two groups were thus hit the hardest by proration, the Gini coefficient was much higher when proration was accounted for in the equity calculations.

It appears that part of the reason proration was so high under Senate Bill 351 was that the state did not accurately forecast the tax increases many districts implemented to take advantage of the second-tier wealth guarantees. But a possibly greater effect was due to the zero-sum nature of the situation facing districts when it became apparent soon after passage of Senate Bill 351 that proration would occur, even during the first year. A district expecting to incur proration of its state entitlement was inclined to increase its tax rate enough to retain the previously announced level of state aid after proration. Districts failing to act in this manner, while others did, would get even less. Therefore, there tended to be a tax rate scramble to stay ahead of proration. With the larger proration levels forecast for 1992-1993, the second year of the biennium, this behavior was even more pronounced that year. The result was a proration of some $609 million in 1992-1993, with the least effect felt by those who raised their tax rates the most in anticipation.

To solve this problem, Senate Bill 7 limited the tax rate guarantee for each district in 1993-1994 to the total tax rate that was levied in 1992-1993. Since both the foundation basic allotment and the guaranteed yield were lower in 1993-1994 than in 1992-1993, many districts lost funds unless they increased their unequalized tax rate. This feature of Senate Bill 7 also had the effect of limiting the state's "liability" for the guaranteed yield, and made it possible for each district to know exactly how much state aid it would receive. Any shortages in state appropriations were made clear to districts at the beginning of the fiscal year rather than at the end as happened under Senate Bill 351 in both 1991-1992 and 1992-1993.

Another important part of Senate Bill 7 was the reduction in funding guarantees established for the foundation (first tier) and guaranteed yield (second tier) portions of the funding formula. Although the legislature appropriated $1 billion in additional funding for the 1993-1995 biennium, this was less than would have been required to continue full implementation of Senate Bill 351. Moreover, the last $250 million was delayed so that the payment was not be made to school districts until the beginning of the 1995-1997 biennium.

To deal with this reduction, the first-tier foundation guarantee, which was to have been $2,400 per pupil in 1993-1994, was reduced to $2,300. Districts are required to levy a tax rate of $0.86 to qualify for this level of funding. The second-tier guaranteed yield was reduced from $22.50 per pupil for each penny of tax effort to $20.55. Moreover, the guarantee is only effective up to the tax rate levied in 1992-1993 by each school district. Thus, most districts that levied the same tax rate in 1993-1994 as they did in 1992-1993 had less money to spend on the education of the children residing in its boundaries.

Approximately 150 districts actually experienced increases in revenue under Senate Bill 7. One group of districts were those that experienced severe proration in 1992-1993. Without proration their revenues increased in 1993-1994 even though their entitlements under Senate Bill 7 decreased. Another group of districts were those medium to low wealth districts with tax rates above $1.35. These districts received increases in entitlements under Senate Bill 7 since all tax rates up to $1.50, 64 cents above the required first-tier rate of $0.86, are rewarded in the second tier. Since only $1.35 of tax levy was equalized in 1992-1993 under Senate Bill 351, these districts had tax levies (between $1.35 and $1.50) that were not equalized in 1992-1993, but were equalized in 1993-1994. Altogether, approximately 15 percent of the districts received more revenue under Senate Bill 7 despite the reductions in the basic allotment and the guaranteed yield.

For the remaining 85 percent of the districts who lost funds under Senate Bill 7, the only way to recoup this loss was to raise local property taxes. However, those increases were not to be equalized until the 1995-1996 biennium under the system devised to eliminate proration. In 1993-1994 districts were allowed to raise the tax rate as much as necessary to regain the per pupil revenue they received in 1992-1993. However, for 1995-1995 any property tax increase exceeding 6 percent was subject to an automatic property tax rollback election. Although there is some confusion regarding legislative intent in passing Senate

Bill 7, Judge McCown in his *Edgewood IV*[13] ruling has stated that the measure places a cap on property tax rates for school district operations and maintenance of $1.50 per hundred of assessed valuation. At the present time, it is not clear what will happen to districts that levy property taxes in excess of $1.50 per hundred.

The result of this funding scheme is that the high wealth districts were forced to raise property taxes substantially to maintain spending levels. This is a direct result of the reductions in their property wealth per pupil mandated by Senate Bill 7. In addition, poor districts that did not want to reduce per pupil spending also had to raise local property taxes since the previous year's tax rate no longer generated the same level of revenue. Given that this additional tax burden was unequalized until the 1995-1997 biennium, those poor districts that did not benefit from the transfers of property from the wealthy districts had to consider substantial local tax increases just to remain even.

Edgewood IV

Despite the complexity and unpopularity of Senate Bill 7, the State Supreme Court upheld the provisions of that law in a ruling on January 30, 1995. Holding that equity is only necessary up to that level where a general diffusion of knowledge is available, it stated that level is an accredited program as established by Texas law. The court states that the district court determined that spending of $3,500 per weighted pupil would meet this requirement. The court further stated that once this level of spending is achieved at a reasonable level of equity (a somewhat lower standard that that of *Edgewood I* and *Edgewood II*), unequalized revenue is permitted. The court found that since the disparity in tax effort between rich and poor districts to achieve a spending level of $3,500 was not great (an average of $1.31 for poor districts and $1.22 for wealthy districts) the mandate of substantially equal access was obtained.

Plaintiffs had argued that since the maximum wealth permitted under Senate Bill 7 was $280,000 per weighted ADA (WADA), and that the guarantee in the second tier was only $20.50 per penny of tax effort, districts whose wealth was less than $280,000 would receive lower revenue than their more wealthy counterparts. For any district with a per pupil wealth of $205,000 or less, the difference at a tax rate of $1.50 is $600 per pupil.[14] The court rejected this argument pointing out that all districts were able to raise $3,500 with tax rates of between $1.22 and $1.31, and accepted that fact that poor districts would have to levy a slightly higher tax rate to reach that revenue level. As for additional local enrichment, the court prohibited districts from levying tax rates in excess of $1.50 for operations and maintenance (general fund) purposes.

Thus spared of having to rewrite the school finance laws yet again, the 1995 session of the Texas Legislature was able to focus on rewriting most other parts of the state's education code.

HISTORICAL ANALYSIS OF THE EQUITY TEXAS SCHOOL FINANCE

Analysis of the Changes in the School Finance Formula

The discussion earlier shows that with the exception of the recapture provisions of Senate Bill 351, the general format of the Texas school finance system did not undergo substantial modifications over time, although there were a number of important changes to the parameters within that structure. Table 9.1 summarizes those changes. Column 1 lists the major legislation responsible for establishment of the parameters summarized in each row. There is no entry for 1988-1989 as that serves as the base year for this study.

Column 2 displays the basic allotment, or the funding guarantee contained in the first-tier foundation program. The data show that from 1988-1989 through 1992-1993 there was a steady increase in this basic allotment, but, as indicated earlier, in 1993-1994 the basic allotment was reduced by $100 per pupil. The third column shows the tax rate that districts were required to levy to receive the basic allotment. For 1988-1989 and 1989-1990 the first-tier tax rate shown in this column was estimated on the basis of a prior calculation, determined by a "local share" parameter established in the statute, and is estimated in Table 9.1. For 1990-1991 and subsequent years the first-tier tax rate was made explicit in the statutes governing the distribution of funds to schools.

Column 3 of Table 9.1 shows that the first-tier tax rate increased substantially, from 31.5 cents per hundred dollars of assessed value in 1988-1989 to 86 cents per hundred in 1993-1994, an increase of 173 percent. Thus, a substantial portion of the increase in total revenues received by school districts was financed through increased property tax levies. However, a portion of this increase can also be attributed to the fact that property wealth per ADA decreased from approximately $231,000 in 1988-1989 to $194,878 in 1992-1993.

Column 4 displays the yield per penny of tax effort guaranteed in the second tier. This is expressed as a yield per penny per pupil in weighted average daily attendance or WADA. WADA is a pupil count based on average daily attendance, with additional weights applied to students with special needs, such as children with disabilities, children who do not speak English, or children who come from low-income homes. Column 5 presents the maximum tax rate that is equalized in the second tier, and column 6 shows the maximum rewarded tax rate for a school district. Districts do not receive state equalization assistance for any taxes levied above the maximum rates displayed in column 6.

Column 7 of Table 9.1 indicates whether or not there is a recapture provision in the first-tier foundation program. Despite the fact that the ruling in *Love* v. *Dallas* precludes the use of statewide recapture, the legislature established the County Education District (CED) system under Senate Bill 351 to provide equalization within counties. Since all of the CEDs had property wealth lower

Table 9.1. Key Parameters and Elements of Texas School Finance Formulas: 1988-1989 to 1993-1994

Year	Primary Legislation (1)	Basic Allotment ($ per pupil) (2)	First-Tier Tax Rate ($ per hundred) (3)	Second-Tier Guaranteed Yield per Penny per Pupil ($)[2] (4)	Second-Tier Tax Effort Limit ($ per hundred) (5)	Maximum Rewarded Tax Rate ($ per hundred) (6)	1st Tier Recapture (Y/N) (7)
1988-1989		1,350	0.315[1]	18.87[1]	0.26[1]	0.575[1]	N
1989-1990	SB 1019	1,477	0.340[1]	18.25	0.36	0.700	N
1990-1991	SB 1	1,910	0.540	17.90	0.37	0.910	N
1991-1992	SB 351	2,200	0.720	21.50	0.45	1.170	Y
1992-1993	SB 351	2,400	0.820	22.50	0.45	1.270	Y
1993-1994	SB 7	2,300	0.860	20.55	0.64	1.500	N[3]

Notes: 1. Estimated.
2. Column 3 refers to WADA, the notion of weighted students used in the second tier.
3. Adjustments to district wealth beginning in 1993-1994 eliminated the need for first tier recapture to achieve a similar or greater improvement in equity. Senate Bill 7 contained provisions which effectively reduced the wealth of districts to the point where recapture began under Senate Bill 351, the previous year. In Senate Bill 351 terms, this wealth reduction had the effect of extending recapture to the second tier as well as the first tier, for the wealthy districts.

Source: Picus and Toenjes, 1994.

than the wealth guarantee of the foundation program, state equalization funds flowed to each CED, even if some of the districts in the CED had to share revenue with other districts in their county or CED. Toenjes (1991) points out that, as a result, the CED equalization system was identical to a statewide recapture system.

Table 9.1 indicates that over the six years reviewed there were dramatic changes in the equalized tax rate, and thus the amount of revenue available to local school districts. This fact is illustrated more dramatically in Table 9.2 which displays the maximum equalized guarantee of each of the school finance systems used between 1988-1989 and 1990-1991. Column 1 of Table 9.2 shows how the total cost of the first tier has increased over the six years studied, and how the first-tier guarantee per ADA has increased (column 3). It is important to note that the figures displayed in column 3 differ from the basic allotment figures in column 2 of Table 9.1, because the basic allotment is based on weighted ADA (WADA) and not ADA. The figures in column 3 of Table 9.2 have been adjusted to account for ADA only. Since there have been very few changes in the weights used for these pupils, and the changes that were made do not appear to have had substantial redistribute effects in terms of money available to school districts, Table 9.2 and the equity analysis that follows converts all figures to ADA to simplify and clarify the analysis.

Table 9.2. Components of Texas School Finance Formulas, 1988-1989 to 1993-1994

School Year	*Total First Tier Cost (millions Of $)* (1)	*Average Daily Attendance* (2)	*Statewide Average First Tier Guarantee ($/ADA)* (3)	*First Tier Tax Rate ($/100)* (4)	*First Tier Wealth Guarantee ($/ADA)*[2] (5)	*Second Tier Wealth Guarantee ($/ADA)*[2] (6)	*Maximum Second Tier Tax Rate Increment ($/100)* (7)	*Maximum Second Tier Grant ($/ADA)* (8)	*Maximum Equalized Gurantee ($/ADA)* (9)
1988-1989	$6,366.2	3,021,753	2,107	0.315[1]	668,822	245,310	0.26	638	2,745
1989-1990	$6,512.4	3,072,253	2,120	0.340[1]	623,455	237,250	0.36	854	2,974
1990-1991	$8,291.9	3,066,161	2,704	0.540	500,801	232,700	0.37	861	3,565
1991-1992	$9,875.0	3,161,405	3,124	0.720	433,835	279,500	0.45	1,258	4,382
1992-1993	$11,210.5	3,257,953	3,441	0.820	419,630	292,500	0.45	1,316	4,757
1993-1994	$11,008.0	3,322,453	3,313	0.860	385,258	267,150	0.64	1,710	5,023

Notes: 1. Estimated.
2. Estimated, converted to average daily attendance pupil basis.

Column 5 offers evidence that despite the dramatic growth in the first-tier guarantee, the wealth level supported through that tier has declined considerably over time, while the wealth guarantee in the second tier has increased somewhat over time. The reductions in wealth guarantees in the first tier tended to reduce state funds going to the wealthiest districts. The increases in the second-tier guarantees, and the higher tax rates rewarded, shifted more state dollars to the lower wealth districts. The combined effects of these two types of changes tended to force up the level of revenues going to former low-spending, low wealth districts, reducing the variation in expenditures per pupil as well as improving the wealth neutrality of the system. The last column in Table 9.2 shows the maximum equalized guarantee per ADA over the six years. The substantial increase in this figure is indicative of the improvements in general that have occurred in terms of school finance equity.

The data displayed in Tables 1 and 2 show how the parameters of the school finance system have changed in the last six years. They do not provide a solid picture of how the equity of the school funding system has changed over that time. As indicated earlier, Verstegen (1987) and Picus and Hertert (1993a) offer evidence that the equity of the system has improved over time. The equity of the school finance system is analyzed below.

School Finance Equity in Texas

This analysis of school finance equity relies on the Berne and Stiefel (1984) equity framework. The Texas data are from the *TEA Snapshot* database published annually by the Texas Education Agency and available in both print and electronic format. For this analysis, the following data elements for each school district were extracted from the *Snapshot* files:

- *Total Enrollment:* Based on fall enrollments, this figure was used for comparative purposes. In most school finance equity analyses adjustments are made for different student needs. Texas has an extensive system of pupil weights for disabled children who are eligible for special education, as well as weights for compensatory education. There are also weights for vocational education (1.37) and for gifted and talented programs (.12 up to 5% of the students). There is even a district weight to account for cost of living differences across school districts. School funds are distributed on the basis of weighted ADA (WADA), rather than enrollment. Different pupil counts are used in the first and second tiers of the finance formula. The first-tier foundation program amount is calculated by applying student weights to an adjusted basic allotment. For tier two WADA is the basis for determining a district's allotment. Additionally, there is a cost index that is applied to the first tier, and half of the index is used for the second tier. Moreover, during the period studied there were very few changes to these weights. Consequently, we chose to conduct this analysis without the added complexity of the pupil weighting scheme. There is no apparent difference in the results of our analysis, and using a consistent pupil count throughout the time period studied emphasizes the equity consequences of the changes in the funding formula. Finally, it should be noted that this analysis was based on enrollments rather than ADA. Since enrollments are higher than ADA, there is some deflation in the level of per pupil spending reported here compared to other sources. However, the numbers reported here match the state's *Snapshot* publication, which is the major dissemination avenue for such data.

- *Taxable Value per Pupil:* Property valuations, adjusted to equalize for variations in local appraisal practices.
- *Equalized Total Tax Rate*: Obtained by the TEA by dividing self-reported levies by the equalized taxable value.
- *State Aid and Local Taxes per Pupil:* The sum of state grants from the foundation program plus local property tax revenues, divided by total enrollments. While certain revenues are not included, such as state contributions to teachers' retirement and local nontax sources, this state foundation grant and local tax total is the notion most often used in assessing the equity status of Texas's system. Furthermore, these are the revenue concepts used in the simulations discussed elsewhere in this paper.

For 1988-1989 the *Snapshot* publication did not show a single state aid plus local tax amount for each district. However, the state amount per pupil was reported, as were the property wealth and tax rates. The local tax portion was calculated from these figures. Another adjustment required was for the revenues for the Houston Independent School District for the year 1991-1992. This was a crucial figure, and erroneous data were reported on the self-report that the TEA received from the district. The necessary figures were estimated from wealth and tax rate information obtained from the TEA.

Table 9.3 contains the equity measures calculated from the *Snapshot* data for the years 1988-1989 through 1992-1993. The table shows that using traditional measures of dispersion, Texas has made considerable progress in recent years toward reducing the variation in per pupil spending across its 1,044 school districts. Mean per pupil expenditures have increased by nearly $1,000 per pupil over the five years analyzed, while the standard deviation declined from $627 per pupil to $563 per pupil. Perhaps more importantly, the coefficient of variation, which is the standard deviation divided by the mean and represents a measure of the "spread" of the data, declined from 0.188 to 0.133 over that period of time.

Another way to analyze spending variations is through the Gini coefficient. The Gini coefficient measures the degree to which the distribution of revenue is perfectly distributed. If each pupil in the state received the same amount of revenue, the value of the Gini coefficient would be zero. The higher the value for the Gini coefficient, the greater the share of total state revenues available to students in high spending districts. As Table 9.3 shows, the Gini coefficient declines from 0.091 in 1988-1989 to 0.058 by 1992-1993. Although this means revenues are not uniformly distributed among students in the state, the declining value for the Gini coefficient indicates that the share of total revenues available to students in the wealthiest districts is declining compared to the share available to students in the poorest districts. The Thiel measure decreases over time, approaching zero by 1992-1993, indicating that the probability a student will receive equal amounts of revenue is increasing over time in Texas.

The McLoone index assesses the equity of the distribution for the lowest spending half of the student distribution. The McLoone index compares the total amount spent for all students below the median per pupil revenue level with a hypothetical calculation of what would have to be spent if all of those students received the median revenue. The closer this value is to one, the less dispersion there is in per pupil revenues among students in low-spending districts. Table 9.3 shows that the McLoone index exceeded 0.9 for all years, and, with the exception of 1991-1992, moves closer to 1.0 each year.

Because there are a few districts with unusually high or unusually low per pupil expenditures, it is often helpful to analyze the dispersion across per pupil spending by considering the districts at the 5th and 95th percentiles of per pupil revenue. The Federal Range Ratio, which is defined as the value of the district at the 95th percentile, minus the value of the district at the 5th percentile divided by the value of the district at the 5th percentile is the most common way to measure this dispersion. Its advantage over a simple ratio of the two values is that it is insensitive to inflation. Table 9.3 shows that the Federal Range Ratio declined consistently over the five years analyzed, indicating that the horizontal equity of the system appears to be improving over time.

While the measures of dispersion provide a good estimation of the horizontal equity in the system, the Texas Supreme Court ruled that an efficient system is one that substantially reduces the relationship between per pupil wealth and available revenues per pupil. In other words, the system should be fiscally neutral. Table 9.3 displays three measures of fiscal neutrality to help assess the equity of the Texas school finance system. To measure the relationship between wealth and per pupil revenues, the first measure presented in Table 9.3 is the correlation coefficient between those two variables. The table shows that in the five years analyzed for this paper, the correlation between wealth and revenue declined from 0.75 to 0.51, indicating that the state has made considerable progress in reducing the relationship between these two variables.

Table 9.3. Equity Measures For Texas School Districts

Equity Measure	*1988-1989*	*1989-1990*	*1990-1991*	*1991-1992*	*1992-1993*
Mean Revenue ($ per pupil)	3,338	3,378	3,655	3,949	4,242
Standard Deviation ($ per pupil)	627	579	598	591	563
Coefficient of Variation	0.188	0.171	0.164	0.150	0.133
Gini Coefficient	0.0911	0.0796	0.0719	0.0650	0.0582
Theil Measure	0.0156	0.0128	0.0113	0.0094	0.0075
McCloone Index	0.920	0.939	0.945	0.932	0.948
Federal Range Ratio	1.256	1.251	1.141	1.077	1.041
Correlation Coefficient	0.751	0.729	0.684	0.540	0.506
Weighted Slope	0.00284	0.00255	0.00239	0.00187	0.00171
Elasticity	0.191	0.152	0.111	0.040	0.033

Source: Data derived from TEA's *Snapshot* Publication.

The weighted slope measures the "steepness" of the relationship between wealth and revenue. The flatter the slope, the less effect increases in wealth have on the level of revenue available to a school district. The elasticity is used to measure this relationship in percentage terms, and a review of the last row in Table 9.3 shows that, like the slope, the elasticity has declined considerably over time. The values for elasticity can be interpreted as meaning that in 1988-1989, a district whose property wealth per pupil was 100 percent higher than another district would be expected to have per pupil revenues approximately 20 percent higher. By 1992-1993 a similar increase in wealth would only lead to a 3 percent higher revenues.

Table 9.4. Analysis of Revenue and Tax Rates for Texas School Districts by Pupil Wealth Deciles: 1988-1989 and 1992-1993

Pupil Decile Grouping	*Avg. Wealth per Pupil (1988-1989) ($)*	*1988-1989 Avg. Total Tax Rate ($/100)*	*1992-1993 Avg. Total Tax Rate ($/100)*	*1988-1989 Avg. Revenue per Pupil ($)*	*1992-1993 Avg. Revenue per Pupil ($)*	*Revenue Change per Pupil ($)*
(1)	(2)	(3)	(4)	(5)	(6)	(7)
0 - 9.99	53,678	0.804	1.213	2,807	4,359	1,552
10 - 19.99	85,034	0.858	1.230	2,848	4,255	1,407
20 - 29.99	105,203	0.895	1.219	3,000	4,345	1,345
30 - 39.99	125,404	0.880	1.266	3,069	4,223	1,154
40 - 49.99	146,534	0.936	1.317	3,085	4,141	1,056
50 - 59.99	168,565	0.938	1.328	3,149	4,041	892
60 - 69.99	207,630	0.970	1.379	3,354	4,131	777
70 - 79.99	242,353	0.975	1.376	3,567	4,167	601
80 - 89.99	300,212	0.844	1.313	3,582	4,176	594
90 - 100.00	509,233	0.729	1.263	4,318	4,740	422
Totals	213,547	0.861	1.307	3,336	4,242	906

Table 9.4 offers a slightly different way of looking at these equity issues. It provides, by pupil wealth deciles, an analysis of tax rates and per pupil revenue in both 1988-1989 and 1992-1993. This table shows a dramatic increase in local property taxes for all districts, along with substantially larger gains in per pupil revenue for the poorer districts, an expected impact of equity improvements in the school finance formula.

The Impact of Senate Bill 351

Although the equity measures contained in Table 9.3 show continuous improvement in equity from 1988-1989 to 1992-1993, the changes introduced by Senate Bill 351, which was effective in 1991-1992 and 1992-1993, are particularly interesting. Recall that the emphasis of the Texas Supreme Court is wealth neutrality. This is measured in Table 9.3 through the correlation, slope, and elasticity. Table 9.5 repeats the slope and elasticity measures, and calculates the percentage change from one year to the next. Table 9.5 shows that the largest improvement or decrease in both of these measures occurred in 1991-1992. For both measures, the percentage reduction in that year was more than double the percentage reduction in any other year studied.

A second way to measure the impact of Senate Bill 351 is to consider the impact that measure had on average tax rates of the poorest and wealthiest districts. The last column of Table 9.6 shows the ratio of the average tax rate in the five poorest deciles to the tax rate in the wealthiest decile. Where the ratio is greater than 1.0, the tax rates in the lower wealth districts exceeded, on average, those in the wealthiest districts. As Table 9.6 shows, in the first three years of the analysis, the average tax rate in the poor half of the state's school districts was approximately 20 percent greater than the average tax rate in the wealthiest decile. However, in 1991-1992, the first year Senate Bill 351 was implemented, difference disappeared, and by 1992-1993 the average tax rate in the wealthiest districts was slightly higher than the average tax rate in the poorest five deciles of districts.

Table 9.5. Texas School Finance Variables: Weighted Slopes, Elasticities and Their Percentage Change, 1988-1989 to 1992-1993

Year	*Weighted Slope*	*Percent Change (%)*	*Elasticity*	*Percent Change (%)*
1988-1989	0.00284		0.191	
1989-1990	0.00255	-10.2	0.152	-20.4
1990-1991	0.00239	-6.3	0.111	-27.0
1991-1992	0.00187	-21.8	0.040	-64.0
1992-1993	0.00171	-8.6	0.033	-17.5

***Table* 9.6.** Comparison of Average Tax Rates in the Wealthiest Decile to the Average in the Five Least Wealthy Deciles: 1988-1989 to 1992-1993

Year	*Wealthiest Decile*	*Lowest Five Deciles*	*Ratio Lowest/Highest*
1988-1989	0.729	0.875	1.20
1989-1990	0.813	0.964	1.19
1990-1991	0.872	1.068	1.22
1991-1992	1.153	1.171	1.02
1992-1993	1.263	1.249	0.99

It appears that the wealthy districts were forced to raise their tax rates considerably to make up for the funds that were lost through recapture. On the other hand, the increased basic allotment in the first tier, and the higher wealth guarantee of the second tier, led to much slower growth in tax rates in the poorest districts in the state. This occurred despite the fact that many of the poor districts increased their second-tier tax rates to take advantage of the large sums of state aid that would follow each penny of tax effort they made.

Summary

This analysis indicates that considerable progress toward both school finance equity, and the Supreme Court's requirement that the relationship between property wealth and per pupil revenues be substantially eliminated have been achieved. However, as indicated earlier, it was not clear that in its initial implementation, Senate Bill 7 would maintain the levels of equity previously established. The next section of the paper compares simulated results of the Texas school finance system under Senate Bill 7 with the distribution of revenues in 1992-1993.

While the data presented previously show dramatic gains in school finance equity through Senate Bill 351, it is helpful to compare the equity achieved through 1992-1993 with that achieved the following year under Senate Bill 7. The next section of this paper makes that comparison, and considers a number of possible adjustments to the school finance data to enhance the value of the equity analysis.

SENATE BILL 7 AND SCHOOL FINANCE EQUITY

Although the statistical measures used to compare the equity of school finance systems have not changed substantially since the publication of Berne and Stiefel's (1984) seminal work in this area, development of accurate cost of education indexes (CEI) has proven to be costly and thus problematic. Moreover, measures of vertical equity have been difficult to develop. One question that needs to be asked in considering these issues is whether adjustments for these factors made much difference in the results of the equity analysis.

Texas provides an excellent opportunity to consider these issues as it has both a weighted pupil program to address the needs of students with special needs, and it incorporates a CEI in the funding distribution formula. Adjustments for both factors were considered in the equity analysis that follows.

Cost of Education Indexes

One of the difficulties in measuring school finance equity is accounting for differences in the cost of providing educational services across a state, a particularly important issue in a large and diverse state like Texas. While a number of methodologies have been developed to measure differences in the cost of education, there have been relatively few states that are willing to incorporate such cost adjustments in their school finance statutes. Texas is one of the few states that has such an adjustment.[15] Even in Texas, the cost index is only used for the first tier of the funding program and to weight half of the second-tier pupil count (WADA).

In the analyses that follow, three different cost index strategies are considered. First, equity results are presented using the index as it is used in Texas. Second, the Texas index is eliminated completely to see what impact leaving out such adjustments has on the overall equity of the system. Finally, since many states do have additional adjustments for small school districts (Texas included), an index that adjusts for school size is used to modify the non-CEI data. Details of each index are provided below.

As indicated earlier, Texas uses a CEI to adjust for differences in the costs of providing educational services across the state. According to Monk and Walker (1991), this index adjusts for both cost differences and provides additional funds for very small school districts. The CEI is used in the distribution of aid to school districts, however, for the guaranteed yield in the second tier, the index is applied to only half of the pupil count used to establish the guarantee. In the results presented below, the first analysis simply uses the CEI as administered by the State of Texas. In the second set of results, this cost adjustment is eliminated, and the equity measured in terms of unadjusted dollars.

The third adjustment considered relies on a national model estimated by Hertert as reported in Hertert, Busch, and Odden (1994). The adjustment is based on the expenditures of school districts from the National Center for Education Statistics Common Core of Data (CCD), 1990.[16] The adjustment provides additional funding for school districts with enrollments of less than 500 students.

Student Counts

Odden and Picus (1992) point out that measures of vertical equity are difficult to develop. One approach is to rely on horizontal measures and weight pupils with special needs. Since Texas has a weighted pupil funding system, estimates of equity based on weighted average daily attendance (WADA) theoretically make that adjustment. In addition to presenting equity findings using WADA, we have also estimated the school finance equity simply using the total number of students in the district. Both options are presented in the analysis that follows.

Estimates of Texas School Finance Equity: 1992-1993 and 1993-1994

To understand the implications of changes in Texas's school finance formula, the following analysis compares the equity of the finance system in 1992-1993 with measures of equity for 1993-1994, the first year Senate Bill 7 was in effect. The equity statistics presented in the tables that follow are the same as in Table 9.3 so that comparisons over time can be made. The data for these analyses come from the Texas Education Agency's (TEA) Public Education Information Management System (PEIMS), and the agency's annual *Snapshot* publications. All were provided in electronic format by TEA.

Table 9.7 presents equity statistics for 1992-1993 and 1993-1994 for state and local revenues on the basis of weighted pupils (WADA). The columns of the table labeled "Texas CEI" present the equity statistics for the actual distribution of state and local revenues in Texas for the respective years. The columns labeled "No CEI" show the equity statistics calculated under the assumption that cost of education differences as defined by the State of Texas are not used. Because the CEI in Texas uses 1.0 as the base and provides districts additional funds for higher costs, the lower average revenue shown when the CEI is eliminated was expected. The columns labeled "Small Dist. Adj." use the revenues estimated without the CEI, and include the small district adjustment discussed previously.

What is interesting about the 1992-1993 data presented in Table 9.7 is that there is very little change in equity of the funding system regardless the assumptions made in the analysis. However, this statement does not hold up in 1993-1994, after the implementation of Senate Bill 7. The equity of the distribution of funds to school districts appears to suffer in comparison to 1992-1993 when Senate Bill 351 was operative. This "step backwards" in equity is somewhat surprising, and most likely explained by two factors: the number of districts that raised unequalized yields to maintain spending levels, and the three-year hold harmless provision that allowed wealthy districts to raise taxes above the cap for three years to regain some of the revenue the property transfer cost them.

Table 9.7. Equity Statistics for Texas School Districts using WADA and State and Local Revenue: 1992-1993 and 1993-1994

	1992-1993			*1993-1994*		
Equity Measure	*Texas CEI*	*No CEI*	*Small Dist. Adj.*	*Texas CEI*	*No CEI*	*Small Dist. Adj.*
Mean Revenue ($ per pupil)	3,792	3,389	3,366	3,970	3,536	4,004
Standard Deviation ($ per pupil)	515	484	409	678	637	913
Coefficient of Variation	0.1356	0.1428	0.1218	0.1707	0.1802	0.2282
Gini Coefficient	0.060	0.060	0.056	0.072	0.073	0.079
Theil Measure	0.000313	0.000319	0.000144	0.00602	0.00630	0.0091
McCloone Index	0.940	0.954	0.953	0.934	0.935	0.935
Federal Range Ratio	1.028	1.078	0.733	1.129	1.183	2.258
Correlation Coefficient	0.522	0.489	0.472	0.491	0.485	0.446
Slope	0.0018	0.0016	0.0013	0.0022	0.00209	0.00276
Elasticity	0.0745	0.0735	0.0605	0.0900	0.0939	0.1095

Another interesting finding from Table 9.7 is the dramatic decline in equity when the small school district adjustment is applied. This adjustment only raises revenues for those districts with fewer than 500 students. Yet apparently it provided enough revenue to a sufficient number of small districts to boost average per pupil spending considerably. Some of these districts were already relatively high spending districts, causing equity to suffer under this scenario.

Table 9.8 offers the same analysis as Table 9.7, only the district's revenues are distributed among the total number of pupils, not the weighted average daily attendance. This makes the figures comparable to the figures displayed in Table 9.3. This adjustment did little to change the patterns observed in the discussion of Table 9.7.

Table 9.8. Equity Statistics for Texas School Districts using Number of Pupils and State and Local Revenue: 1992-1993 and 1993-1994

	1992-1993			*1993-1994*		
Equity Measure	*Texas CEI*	*No CEI*	*Small Dist. Adj.*	*Texas CEI*	*No CEI*	*Small Dist. Adj.*
Mean Revenue ($ per pupil)	4,242	3,790	3,765	4,433	3,948	4,471
Standard Deviation ($ per pupil)	563	543	461	733	700	1,001
Coefficient of Variation	0.1327	0.1433	0.1225	0.1655	0.1773	0.2238
Gini Coefficient	0.058	0.062	0.057	0.070	0.074	0.077
Theil Measure	0.000313	0.000319	0.000144	0.006064	0.006349	0.000166
McCloone Index	0.944	0.945	0.945	0.933	0.927	0.933
Federal Range Ratio	1.041	1.078	0.706	1.130	1.181	2.255
Correlation Coefficient	0.506	0.463	0.445	0.486	0.475	0.439
Slope	0.0017	0.0015	0.0012	0.0022	0.0020	0.0027
Elasticity	0.0711	0.0623	0.0577	0.0867	0.0908	0.1058

Table 9.9. Equity Statistics for Texas School Districts using WADA and Total Revenue: 1992-1993 and 1993-1994

	1992-1993			*1993-1994*		
Equity Measure	*Texas CEI*	*No CEI*	*Small Dist. Adj.*	*Texas CEI*	*No CEI*	*Small Dist. Adj.*
Mean Revenue ($ per pupil)	4,243	3,790	3,766	4,311	3,839	3,872
Standard Deviation ($ per pupil)	581	537	461	691	644	880
Coefficient of Variation	0.1370	0.1418	0.1223	0.1603	0.1678	0.2271
Gini Coefficient	0.063	0.063	0.060	0.067	0.067	0.075
Theil Measure	0.000289	0.000295	0.000133	0.00556	0.005831	0.008838
McCloone Index	0.938	0.935	0.935	0.925	0.937	0.937
Federal Range Ratio	1.126	1.122	0.824	1.081	1.107	2.3514
Correlation Coefficient	0.438	0.417	0.388	0.430	0.432	0.397
Slope	0.0017	0.0015	0.0012	0.0020	0.0019	0.0024
Elasticity	0.0631	0.0623	0.0500	0.0741	0.7786	0.0969

Picus's (1994) study of the impact of school finance reform on local school districts in Texas found that some poor districts in the state not only received dramatic increases in per pupil revenues, but that they already had tremendous resources from the federal government. In theory, if the federal government were seeking to provide categorical assistance to the most needy students, the impact of federal funds would be to improve school finance equity (if it could be shown that poor children lived in places with low property wealth). Tables 9.9 and 9.10 address this question by providing the same equity measures as reported in Tables 9.7 and 9.8, only in place of state and local revenues, total district revenues are used. Again, perhaps the most significant thing to observe from

Table 9.10. Equity Statistics for Texas School Districts using Number of Pupils and Total Revenue: 1992-1993 and 1993-1994

	1992-1993			*1993-1994*		
Equity Measure	*Texas CEI*	*No CEI*	*Small Dist. Adj.*	*Texas CEI*	*No CEI*	*Small Dist. Adj.*
Mean Revenue ($ per pupil)	4,746	4,239	4,212	4,816	4,286	4,324
Standard Deviation ($ per pupil)	627	594	509	752	708	972
Coefficient of Variation	0.1320	0.1401	0.1208	0.1562	0.1652	0.2248
Gini Coefficient	0.061	0.064	0.060	0.066	0.068	0.076
Theil Measure	0.000289	0.000295	0.000132	0.005601	0.0059	0.000161
McCloone Index	0.936	0.930	0.928	0.934	0.934	0.934
Federal Range Ratio	1.068	1.069	0.795	1.073	1.113	2.272
Correlation Coefficient	0.426	0.398	0.368	0.418	0.421	0.387
Slope	0.0016	0.0014	0.0011	0.0019	0.0018	0.0023
Elasticity	0.0596	0.0591	0.0472	0.704	0.0750	0.0938

these is how similar the equity statistics continue to be. Again, most measures of equity moved in the "wrong direction" in 1993-1994 as explained earlier. However, the real question is what will happen to equity across the system when Senate Bill 7 is fully implemented and the wealthiest districts are limited to a taxable value per pupil of $280,000. The next section of the paper offers some preliminary answers to this question.

SIMULATING SCHOOL FINANCE EQUITY FOR 1996-1997

To estimate the full effects of Senate Bill 7 a number of simulations were run for 1996-1997, the first year in which the requirements of that law will be fully implemented. Since the equity statistics that resulted were again very similar across alternatives, only two are presented in Table 9.11. The first simulation simply estimated the distribution of funds to school districts under Senate Bill 7 as it was drafted and passed by the legislature. The second simulation displayed in Table 9.11 assumed that the guaranteed yield of the second tier would be set at $280,000, the maximum wealth permitted.

In theory, if all of the districts in the state set their tax rate at $1.50, then there would be perfect horizontal equity for state and local revenue across all of them. Since it is unlikely that will happen, the simulation increased tax rates enough for 1993-1994 to make up for any revenue loses associated with the lower funding guarantees of Senate Bill 7, and then allowed tax rates to increase by no more than six cents per year, to correspond with the maximum increases districts could implement without facing an automatic rollback election.[17]

Table 9.11. Equity Statistics for Texas School Districts using Total Pupils and State and Local Revenue: Estimated 1996-1997

Equity Measure	*SB 7*	*Second Tier at $280,000*
Mean Revenue ($ per pupil)	3,590	3,948
Standard Deviation ($ per pupil)	349	396
Coefficient of Variation	0.0967	0.103
Gini Coefficient	0.053	0.056
Theil Measure	0.000030	0.000023
McCloone Index	0.932	0.918
Federal Range Ratio	0.556	0.552
Correlation Coefficient	0.429	0.263
Slope	0.0022	0.0016
Elasticity	0.8023	0.0511

Table 9.11 shows dramatic improvements in the equity measures compared to the data for 1993-1994 described earlier. The most important reason for this is the dramatic reduction in revenues for the wealthiest and high-spending districts. For example, in the 1993-1994 estimates the district with the highest per pupil revenue received over $19,000, and even at the 95th percentile, total state and local revenue amounted to $8,533 per pupil. However, in the 1996-1997 estimates, under the provisions of Senate Bill 7, the highest revenue recorded was $4,831, and the observation at the 95th percentile was $4,358 per pupil. The dramatic declines in the range are confirmed by the substantial drop in the Federal Range Ratio displayed in both columns of Table 9.11. Other measures of equity showed similar and equally dramatic declines.

CONCLUSIONS

The Texas Supreme Court's ruling in *Edgewood IV* has both ended the current round of school finance litigation in the Lone Star State, and may have precluded a number of potential future law suits as well. As Table 9.11 shows, once fully implemented, the wealth restrictions of Senate Bill 7 will dramatically limit the revenue potential of the wealthiest districts, forcing dramatic reductions in spending for those districts. Earlier rulings in *Edgewood I* and *II* seemed to call for such dramatic equalization, and it is not surprising that the way the legislature elected to reach this goal was to level the high spending districts down.

This leveling down might be the source of a new round of law suits focused on the issue of adequacy. However, the court seems to have dealt a blow to such suits in its ruling by stating that a funding level of $3,500 per student is adequate. As Table 9.11 shows, our estimates of per pupil state and local revenues for 1996-1997 average $3,590 per pupil.

One matter the legislature might want to consider in the future is reductions in the $600 funding gap that exists between poor districts (those with wealth of $205,000 per pupil or less) and districts with a wealth of $280,000 per pupil. Our estimates show that even if the second-tier guarantee is increased to the $280,000 level, not all districts will choose to levy the full $1.50 tax rate immediately. Given the high levels of equity measured by this research, Texas clearly offers one of the most equitable funding formulas among the 50 states. Whether additional equity measures are necessary is a matter of policy debate since the Texas Supreme Court has allowed this system to stand.

A look at the equity statistics presented in Tables 9.7-9.11 show that the distribution of funds in Texas is among the most equitable of all the 50 states. Given the history of the *Edgewood* litigation, and the court's apparent desire to be done with this issue, it seems unlikely that a future challenge to the school finance system will succeed in the near future.

ACKNOWLEDGMENT

The research reported in this paper was supported by the Finance Center of the Consortium for Policy Research in Education (CPRE), a consortium of the University of Pennsylvania, the University of Wisconsin-Madison, Harvard University, the University of Michigan, and Stanford University. The work was supported by the U.S. Department of Education, Office of Educational Research and Improvement, Contract RR91172002. The views expressed are those of the author and not necessarily shared by USC, the Center for Research on Education Finance (CREF), CPRE or its partners, or the U.S. Department of Education. I would like to thank Allan Odden, Catherine Clark, and Laurence Tonjes for their helpful comments and review of an earlier draft of this chapter.

NOTES

1. Texas Supreme Court, No. 94-0152, January 30, 1995.
2. For a description of how changes in the school funding formula has impacted various types of districts, see Picus (1994).
3. *Edgewood Indep. Sch. Dist.* v. *Kirby*, No. 362,516 (259th Dist. Ct. Travis Cty. Tex. June 1, 1987), *rev'd* 761 S.W. 2d 859 (Ct. App. Tex. 1988), *rev'd* 777 S.W. 2d 391 (1989) (hereinafter *Edgewood I*).
4. For a detailed discussion of measures of school finance equity, see Berne and Stiefel (1984).
5. *Rodriguez* v. *San Antonio Indep Sch. Dist.*, 337 F. Supp. 280 (W.D. Tex. 1971), rev'd 411 U.S. 1 (1973).
6. *San Antonio Indep. School District* v. *Rodriguez*, 411 U.S. 1 (1973).
7. *Edgewood Indep. Sch. Dist.* v. *Bynum.*
8. *Edgewood I.*
9. *Edgewood II.*
10. *Love* v. *Dallas* 40 S.W.2d 20 (Tex. 1931).
11. *Edgewood III.*
12. The figure of $280,000 was established in law by the legislature and represents the 86th percentile of district per pupil wealth across the state.
13. *Edgewood Indep. Sch. Dist.* v. *Meno*, No. 362,516-B (250th Dist. Ct. Travis Cty. Tex. January 25, 1994).
14. $205,000 represents the 77th percentile of district per pupil property wealth.
15. For details on the construction of the Texas cost of education index, see Monk and Walker (1991). For information on the construction of such indexes generally, see Chambers (1981).
16. To develop the adjustment, 1989-1990 expenditures per pupil for each district were standardized by dividing each district's per pupil expenditures by its respective state average. This ratio was used as the dependent variable, with enrollment as the independent variable. The best model was found to be a quadratic form, using elementary schools as a dummy variable. High school districts were eliminated from the sample because there were so few of them and the data were truncated to include only those districts with an enrollment of 800 or fewer students. The resulting coefficients showed that size was only a factor in districts with 500 or fewer students. The specific adjustment used for unified districts was:

$$\text{Factor} = (2.001\text{-}0.00356\text{ENR}) + (0.000003116\text{ENR}^2)$$

17. Note that in 1995 Senate Bill 1 modified the rollback tax collection to account for enrollment growth and increased the rate limit to eight cents above the effective tax rate.

REFERENCES

Berne, R., & Stiefel, L. (1984). *The measurement of equity in school finance.* Baltimore, MD: The Johns Hopkins University Press.

Chambers, J.C. (1981). Cost and price level adjustments to stte aid for education: A theoretical and empirical view. In K.F. Jordan& N. Cambron-McCabe (Eds.), *Perspectives in state school support programs* (pp. 39-85). Cambridge, MA: Ballinger.

Clark, C. (1993). The Texas school finance melodrama. Paper presented at the annual meeting of the National Tax Association, November 8, Saint Paul, Minnesota.

Hertert, L. Busch, C., & Odden, A. (1994). School financing inequities among the states: The problem from a national perspective. *Journal of education finance, 19* (3), 231-255.

Hobby, W.P., & Walker, B.D. (1991). Legislative reform of the Texas public school finance system, 1973-1991. *Harvard Journal on Legislation, 28* (367), 379-394.

Monk, D.H., & Walker, B.D. (1991). The Texas cost of education index: A broader approach. *Journal of education finance, 17* (2), 172-192.

Odden, A.R., & Picus. L.O. (1992). *School finance: A policy perspective.* New York: McGraw Hill.

Picus, L.O. (1994). The local impact of school finance reform in Texas. *Educational Evaluation and Policy Analysis, 16* (4), 391-404.

Picus, L.O., & Toenjes, L. (1994). Texas school finance: Assessing the impact of multiple reforms. *Journal of Texas Public Education, 2* (1).

Picus, L.O., & Hertert, L. (1993a). A school finance dilemma for Texas: Achieving equity in a time of fiscal constraint. *Texas Researcher, 4,* 1-28.

________. (1993b). Three strikes and you're out: Texas school finance after *Edgewood III. Journal of Education Finance, 18* (4).

Toenjes, L.A. (1991). Structural change to Texas' school finance formula. *Journal of Education Finance, 17,* 224-254.

Verstegen, D.A. (1987). Equity in state education finance: A response to *Rodriguez. Journal of Education Finance, 12,* 315-330.

Walker, B.D. (1993). *Highlights of Senate Bill 7* [*Tax base reduction*]. Austin, TX: Texas Association of School Boards, mimeo.

Chapter 10

EDUCATIONAL ACHIEVEMENT AND TEACHER QUALIFICATIONS:

NEW EVIDENCE FROM MICROLEVEL DATA

Dominic J. Brewer and Dan D. Goldhaber

INTRODUCTION

Public schools in the United States have faced a great deal of criticism over the past three decades. Unflattering international comparisons of student performance and declines in some average standardized test score measures occurred during a period when real per pupil expenditures on K-12 schooling increased dramatically. Many have come to the conclusion that the nation's public schools were failing.[1] The past decade, spurred by a spate of reports in the early 1980s, has seen numerous reforms scattered throughout the country, which have included new forms of student assessment, curriculum innovation, and new classroom organization, as well as changes in school

Advances in Educational Productivity, Volume 6, pages 243-264.

ISBN: 0-7623-0082-5

governance and financing. These reforms have been accompanied by increased research by economists, sociologists, and others seeking to determine the relationship between educational performance and schooling expenditures or other measures of school "quality."

Since the 1966 Coleman report (Coleman et al., 1966), literally hundreds of studies use what has become known as the "educational productivity" paradigm. Most of this research seeks to explain variation in standardized test scores at student, school, or school district level. Typically, multiple regression models are utilized in which individual and family background characteristics and schooling variables such as expenditures per pupil and class size are included as factors explaining test scores. This work concludes that individual and family background traits explain the vast majority of variation in student test scores. The effects of educational inputs such as per pupil spending, teacher experience, and teacher degree level on outcomes, are less certain. Some researchers have noted inconsistent findings across studies (Hanushek, 1986) and concluded that there is no systematic relationship between observable schooling resources and student test scores, at least at current levels of resource utilization. Others interpret the same literature somewhat more positively (Hedges, Laine, & Greenwald, 1994).

The quest to identify the true relationship between educational inputs and outcomes continues; every year additional research studies are published on the issue. Over the past decade great strides have been made in this literature due in part to the availability of more detailed, student-level data representative of the nation. For example, two large-scale longitudinal databases collected by the National Center for Education Statistics (NCES) have followed large cohorts of students through high school in the early 1980s (*High School and Beyond*) and from eighth grade to the twelfth grade in the late 1980s and early 1990s (*National Educational Longitudinal Study of 1988*). Progress has also been made in the development of more sophisticated statistical techniques which have been applied to the productivity issue (for example, the use of "hierarchical linear models"). However, although there is some evidence that newer studies are generating more positive findings in the sense of finding that inputs do matter, firm conclusions remain elusive.

The uncertain results with regard to teachers are particularly puzzling. Few would deny that teachers play a central role in the education of our nation's children. Teaching is the largest profession in the United States, employing over three million adults (NCES, 1994, p. 71). An elaborate system of teacher education and certification is geared toward the preparation of those entering teaching, and there are significant professional development opportunities for those remaining in the profession. More than 40 percent of teachers have at least a Master's degree and more than 25 percent have at least 20 years full-time teaching experience (NCES, 1994, p. 77). Over 60 percent of all schooling expenditures at the K-12 level are devoted to instructional costs which consist overwhelmingly of teacher salaries and benefits (see, for example, Brewer, 1996; Picus & Watenbarger, 1995). Further, teacher salary incentives reward years of experience and degree levels, traits that do not appear to have a relationship to student achievement. What can explain the inconsistent findings of the educational productivity literature with respect to educational resources, particularly teachers? In this chapter we shed some light on this question by exploring the relationship between student achievement and schooling inputs.

First, we review the previous literature on educational productivity and outline some of the potential problems with it. Second, we present some new evidence of our own. We use student-level data from the *National Educational Longitudinal Study of 1988* (NELS), one of the most comprehensive nationally representative data sets available for educational productivity analyses. These data contain information on the background of eighth-grade students in 1988, who were re-surveyed in 1990 and 1992. Unlike most other data, NELS links students to a specific class and teacher, and thus marks a significant improvement over most previous surveys. We model mathematics achievement as a function of individual and family background characteristics for our sample of public school students. We add a familiar set of schooling, teacher, and classroom characteristics to our model. We find that several teacher characteristics—in particular, teachers' math specific preparation—do have a statistically significant effect on student achievement in the expected direction. This corroborates some other recent research on the effects of teacher subject matter preparation.

Following Murnane and Phillips (1981) we broaden the measured teacher characteristics by also focusing on a set of variables describing teacher behavior, such as whether teachers utilize small groups or use oral questioning strategies. We find that these traits explain part of the variation in student achievement, independent of teacher characteristics. We then briefly discuss some further evidence on teachers' effects on schooling outcomes presented by us elsewhere (Goldhaber & Brewer, forthcoming), which uses more sophisticated statistical techniques in order to determine the effect of omitting unobservable teacher traits on the estimated relationship between observable teacher characteristics and achievement. These results suggest that traditional educational production functions may not, despite their problems, have yielded biased estimates of the effects of schooling resources on test scores; the recent, rather more positive results found by us and others, most likely reflect the use of detailed, student-level data.

PREVIOUS LITERATURE ON EDUCATIONAL PRODUCTIVITY

A large literature in education research is devoted to identifying the correlates of successful schooling. This body of work is very diverse, ranging from the so-called "effective schools" paradigm of the early 1980s, which sought to identify a set of factors common to high and low performing schools (Purkey & Smith, 1983; Rosenholtz, 1985), to "process-product" research, which has attempted to link particular pedagogical practices to measures of outcomes (Brophy & Good, 1986; Good, 1994), often at the student level. The "educational productivity" paradigm has, over the past 25 years, produced a set of predominantly empirical studies (conducted by economists) quantifying the relationship between schooling resources and student outcomes in the context of a causal model of the school "production process" in which "inputs" are transformed into "outputs" (Hanushek, 1986). Outcomes have usually been measured by standardized test scores, although in recent years there has been a resurgence of academic interest in the relationship between school "quality" and other outcomes such as wages (Card & Krueger, 1992; Betts, 1995), graduation from high school, and college attendance (Evans & Schwab, 1995).[2] In standard educational productivity models the achievement of student i at school j, Y_{ij}, is regressed on a vector of individual and family background variables (including some measure of prior ability or achievement), X_{ij}, and a vector of schooling resources, S_j, which do not vary across students, and a random error term:

$$Y_{ij} = \beta X_{ij} + \gamma S_j + \epsilon_{ij} \tag{1}$$

S_j may consist of school, teacher, or class specific variables such as school spending per student, the percentage of teachers at a school with a Master's degree or mean years of teacher experience, a teacher's years of experience, degree level and degree major, and class size. β is the return to individual and family background characteristics and γ is the return to schooling resources. If (1) is correctly specified, ordinary least squares (OLS) estimation will yield consistent estimates of β and γ. Hence, in this case, γ precisely quantifies the effect of changing schooling inputs on student test scores. A simple t-test of each estimated γ indicates whether we can infer with any certainty that our estimated resource effects are in fact statistically different from zero. The overall importance of schooling factors S_j can be ascertained by performing an F-test of the hypothesis that the coefficients of the schooling variables are jointly equal to zero.[3]

The notion that there is an estimable education production function for a set of individuals within or across classes or teachers or schools or school districts is not unchallenged (Monk, 1992). Like any model, the education production function is certainly a simplification of reality, but it is a useful tool. This is particularly true for policy purposes because most applications focus on manipulable, measurable inputs rather than on intangible variables or amorphous constructs. One might view the type of multiple regression model in (1) as simply a way to tease out the relationship between schooling resource measures and test scores, rather than inferring any causal relationship between the two. The underlying productivity paradigm does, however, make this inference.

What does the educational productivity literature show? In an often cited article, Eric Hanushek states that "differences in [school] quality do not seem to reflect variations in expenditures, class sizes, or other commonly measured attributes of schools and teachers" (Hanushek, 1986, p. 1142). To back this claim, he cites numerous studies that employ an educational production function methodology and show conflicting results as to the importance of schooling resources. He concludes that there is "no strong evidence that teacher/student ratios, teacher education, or teacher experience have an expected positive effect on student achievement" and that "there appears to be no strong or systematic relationship between school expenditures and student performance" (Hanushek, 1986, p. 1162). An updated article in 1989 reached similar conclusions (Hanushek, 1989). The mixed findings reported by Hanushek, combined with the more robust result that individual and family background characteristics can explain the majority of variation in student test scores, has led some (but not necessarily Hanushek) to jump to the conclusion that schools and teachers may not matter—at least in the sense that additional monies spent on educational resources are wasted.

Of course, within the hundreds of educational productivity studies there are some that have produced "positive" results in the sense that they find educational inputs to be related to student outcomes in the expected direction. This is true of some teacher inputs such as degree level and subject matter preparation (Monk & King, 1994) and particularly true of studies that have examined measures or proxies of teacher ability (Ferguson, 1991; Ehrenberg & Brewer, 1994, 1995). Such studies tend to be cited often by those wishing to increase real resources in public schools; opponents on the other hand can easily find other studies that reach apparently opposite conclusions. This makes interpretation of the literature on educational productivity controversial. In fact, the debate over educational productivity has become a highly politicized one (Boyd & Hartman, 1988). This is more than academic skirmishing; for example, rival studies are used extensively by both sides of school finance equalization court cases. A reasonable interpretation of the literature is that the findings have certainly been inconsistent.

How can these results be explained? If we accept Hanushek's interpretation of the literature and methodology underlying the studies he reviewed, one possibility is that (public) schools have a suboptimal allocation of resources (allocative inefficiency) and/or do not operate on the production possibility frontier (technical inefficiency). In these cases additional teacher inputs or smaller class sizes would not necessarily imply higher output, ceteris paribus.[4] This result does not imply that schooling resources may never affect student achievement positively, simply that given the way public schools are organized, additional resources do not make much systematic difference. This may also be due to lack of variation across school settings in variables like class size, making it difficult to identify the effects of such characteristics.

On the other hand, it may be premature to reach strong conclusions about previous research. It is possible to challenge Hanushek's "consensus" view of the literature. He reached his conclusions by noting the direction of estimated input effects (teacher/pupil ratio, teacher education, teacher experience, teacher salary, per pupil expenditures, administrative inputs, and facilities) on student achievement, along with whether they were statistically significant, and simply tallying ("vote counting") the number of statistically significant positive and negative coefficients. A recent "meta-analysis" by Hedges, Laine, and Greenwald (1994) using the same set of studies reviewed by Hanushek reached a very different conclusion. Their basic argument was that the *pattern* of estimated coefficients in these studies suggested there were indeed systematic positive effects. (A follow-up study using updated research reaches the same conclusion [Laine, Greenwald, & Hedges, 1995].) Since this study has received widespread attention, especially among education researchers, it deserves brief comment.

The basic approach of Hedges, Laine, and Greenwald was to reanalyze the same 38 separate articles and books reviewed by Hanushek (consisting of up to 152 separate models). At its most basic, their argument is that if teacher experience and student achievement were really unrelated, one would expect to find when looking at a large number of studies 50 percent showing a positive relationship and 50 percent a negative one, with only 5 percent of the studies having statistically significant positive or negative findings. The simple vote counting methodology of Hanushek does not reveal this; in fact many more studies produce statistically significant results than one might expect if there was no relationship between inputs and outputs. Hedges, Laine, and Greenwald go on to argue that such a vote counting methodology is flawed and has a tendency not to detect effects even when they are in fact present. The authors instead rely on "meta-analysis," which permit them to perform "combined significance" tests across studies (i.e., combine statistical significance values from different studies to obtain an overall level of significance). They also allow for the fact that Hanushek used several model results from the same study which implies the estimates should not be treated independently, for the disproportionate influence of any one study, and for the effects of "outlier" studies which find very small or very large effects. Based on their analyses, Hedges, Laine, and Greenwald conclude that "the production function studies of the relation between resource inputs and school outcomes examined by Hanushek do not support his conclusion that resource inputs are unrelated to outcomes...reanalysis with more powerful analytic methods suggests strong support for at least some positive effects of resource inputs and little support for the existence of negative effects" (Hedges, Laine, & Greenwald, 1994, p. 13).

In reply, Hanushek (1994) makes a strong case for his original conclusions, pointing out that meta-analysis techniques require a set of assumptions that are not met in the education production function case. Of particular importance is that in order to conduct their analyses Hedges, Laine, and Greenwald "throw out" many studies that find statistically insignificant results—as many as 30-40 percent of the original sample for some important inputs like teacher education. This calls into question the conclusions these authors reach. Overall the salience of this recent challenge to the conventional wisdom of Hanushek hinges on the appropriateness of methods of combining the results from numerous disparate studies. The semantics of the debate unfortunately dominate the important policy implications.

A more compelling argument against accepting Hanushek's conclusion is that many of the studies he reviewed are "bad studies" and should not be given much weight; Hanushek made little attempt to separate studies on the basis of quality save that they had been published. Further, his initial review was essentially completed in the early 1980s and based on studies completed in the previous decade. There are good reasons to believe that many educational production function studies, particularly those completed in the 1970s, had major deficiencies in empirical methodology and available data (these are not mutually exclusive). If the methodological objections are reasonable then an indiscriminate review of the literature will not yield an accurate assessment of the effects of schooling inputs on student outputs.[5]

Data deficiencies imply several problems. The most important of these is that key variables may have been omitted from estimated test score models, potentially leading to biased estimated coefficients of the included variables. For example, many early studies were unable to control for prior achievement using a "pretest" score to net out individual ability, as is now generally accepted to be important (Boardman & Murnane, 1979; Hanushek, 1979; Hedges, Laine, & Greenwald, 1994). Variables representing school and teacher "quality" that are used in most production function studies are typically very crude. For example, teacher degree levels and years of experience may be only weakly related to teaching skill. Degree level alone does not distinguish between colleges of differing quality, nor when the degree was granted, nor does it convey any information about college major, certification requirements fulfilled, or subsequent professional development. Teacher motivation, enthusiasm, and skill at presenting class material are likely to influence students' achievement, but are difficult traits to accurately measure and are thus omitted from standard regression analyses.

The effects of omitted variables bias and various econometric approaches for dealing with the problem are laid out more formally by us in a technical paper (Goldhaber & Brewer, forthcoming). Here, we simply note some of the problems that likely arise in the estimation of education production functions. Assume, for example, that the vector of schooling resources S in (1) above can be decomposed into two parts: observable characteristics, Z_1, such as class size, teacher experience, teacher degree level, or per pupil expenditure, which are included in the model; and unobservable characteristics, Z_2, such as teacher skill, behavior and motivation, and classroom peer effects, which are omitted from the model. The omission of Z_2 may cause two problems. First, the total effect of schooling resources on student outcomes could be understated because omitted factors (Z_2) will not be included in the explained portion of the variation in student achievement. Second, we may obtain biased estimates of the effects of particular schooling resources on student outcomes. Suppose, for example, that children from upper-income families attend schools with more motivated teachers but that teacher motivation is unobservable. In this case, assuming teacher motivation positively impacts student achievement, the estimation of the educational returns to increases in family income will be overstated. Similarly suppose that younger, more inexperienced teachers are the most motivated; here the estimated effects of teacher experience on student achievement would be understated.

The obvious solution to this problem is to utilize data that includes the relevant variables or at least good proxies for them. This might be done by adding measures of teacher behavior or teacher ability, or refined measures of peer effects. Production function studies that have used more refined measures of teacher inputs have found more consistent results. Monk and King (1994) report that teacher subject matter preparation in mathematics and science does have some positive impact on student achievement in those subjects. Measures of the selectivity of teachers' colleges have also been shown to be positively related to student achievement (Ehrenberg & Brewer, 1994). The latter result most likely reflects the fact that the selectivity measure captures teacher ability. The few studies that have had measures of teacher (verbal) ability, for example, in the form of a test score, have found a much more robust positive relationship to student achievement (Coleman et al., 1966; Ehrenberg & Brewer, 1995; Ferguson, 1991) than those using other teacher characteristics.

A further issue with regard to teaching "inputs" is that what actually takes place in the classroom (the process of teaching and learning) may not be captured by teacher characteristics alone (Murnane & Phillips, 1981). This includes both the specifics of the curriculum, the way in which material is conveyed to students by teachers, and the dynamics of student and teacher interaction in any classroom. There is a long tradition among education researchers investigating the relationship between teacher *behavior* and student outcomes (see Brophy & Good, 1986 for a review). Although this "process-product" research often uses a univariate approach (e.g., simple pairwise correlations) and small samples, it has produced some relatively strong results, particularly in regard to the importance of curricula exposure and some pedagogical strategies. If such variables are important predictors of student achievement and are not captured by included school and teacher characteristics, traditional educational production functions may again yield biased results.

While recently available educational data have made it possible to include a broader set of variables describing educational data, it is unlikely that any data set will contain sufficient information. It is also possible that data deficiencies have resulted in significant measurement error problems in previous studies. For instance, it is possible that schooling inputs, particularly class size and expenditures per pupil, are measured with error. Some error likely arises from aggregation of variables to school level.[6] Rather than class size, studies often utilize total school enrollment divided by total number of teachers as an average pupil/teacher ratio. This can lead to dramatically different estimates of the effects of school resources on achievement (see, for instance, Akerhielm, 1995). Of particular concern is that "teacher characteristics" used in many studies are not in fact those of any individual teacher but rather a school-level variable such as mean years of experience, percentage of teachers with at least a Master's degree, or mean teacher salary. Use of these measures ignores the considerable variation in teacher characteristics that may exist within a school, and makes it problematic to accurately access the impact of specific teacher characteristics on achievement.

The conventional view that observable school inputs, and teachers in particular, do not positively impact student achievement rests on somewhat shaky empirical grounds. While Hanushek's original synthesis of the educational productivity literature and his conclusions regarding the effects of school resources on test scores seem reasonable to us, this view was based on older studies. The main problem is likely to be omitted variable bias arising from inadequate data and extremely crude proxies for teacher skill found in such educational production function research. In the absence of more refined measures, it is important to establish whether (i) more comprehensive, student-level data yields similar conclusions, and (ii) the omission of unobservables biases the estimates of the variables that are included in the educational production functions. If the latter is true, then it calls into serious question the usefulness of this approach for public policy.

SOME NEW EVIDENCE

Data

We now present some new evidence on educational productivity, focusing primarily on teacher inputs. We use data taken from the first two waves of the *National Educational Longitudinal Study of 1988* (NELS), a nationally representative survey of about 24,000 eighth-grade students conducted in the spring of 1988. Eighteen thousand of these students were resurveyed in tenth grade (spring 1990). Students provided comprehensive information on themselves and their families (their race/ethnicity, sex, family structure), supplemented by a parental survey in 1988 (providing information on, for example, family income). At the time of each survey students took one or more subject-based tests in mathematics, science, writing, and history. The data, therefore, permit the estimation of "value-added" or gain score production functions that control for previous knowledge and/or ability. We confine our attention to those students who took the mathematics achievement test in the eighth and tenth grade, the tenth-grade score being modeled as a function of the eighth-grade score and other variables.[7] The tests were carefully designed by the Educational Testing Service to assess mathematics knowledge. While all students took the same test in the eighth grade, tests of varying difficulty were given to students in the tenth grade depending on their eighth-grade score in order to guard against "ceiling" and "floor" effects. The tests were given a common scale using Item Response Theory (IRT).[8]

The unique feature of NELS is that it provides detailed teacher- and class-level information that is tied directly to individual students. In other words, the characteristics of each tenth-grade mathematics teacher (sex, race/ethnicity, degree level, experience, certification, etc.) who taught students taking the tenth-grade mathematics test is known. In addition, some characteristics of the classes one teacher taught are also available (e.g., class size, track class composition).[9] This feature of the data makes NELS one of the most comprehensive and well-designed studies available for conducting educational production function analyses. School administrators provided information on their schools: for example, in regard to the composition of the teaching staff (percentage with a Master's degree, percentage white, etc.) and school quality indicators (e.g., percentage of the school's graduates going on to four-year colleges).

We confine our attention to a sample of 5,149 tenth-grade public school students for whom complete data are available. Public school students are examined because of the difficulties of controlling for the nonrandom assignment of students to public and private schools (Goldhaber, 1996). Further, any policy conclusions resulting from our analyses apply primarily to public schools. These students are drawn from 638 schools, with 2,245 individual mathematics teachers. The sampling frame of NELS is such that there are not only multiple students per school and per teacher, but there are also multiple classes per teacher; there are 3,498 classes in our sample. This is also a unique advantage of the NELS data which, at least in principle, permits the estimation of a broader class of models than is usually the case. Means of all variables used in our analyses are shown in Table 10.1.

Table 10.1. Sample from the National Educational Longitudinal Study of 1988: Variable Means (Standard Deviations)

Category	*Mean (Standard Deviation)*	
Individual and Family Background Variables		
10th grade mathematics test score	43.94	(13.6)
8th grade mathematics test score	36.57	(11.7)
Female	0.51	(0.5)
Black	0.09	(0.3)
Hispanic	0.11	(0.3)
Asian	0.05	(0.2)
Parental education	14.07	(2.3)
Family income	37152.48	(31418.82)
Mother only	0.07	(0.3)
Father only	0.14	(0.1)
School Variables		
Urban	0.20	(0.4)
Rural	0.41	(0.5)
Northeast	0.15	(0.4)
North central	0.33	(0.5)
West	0.17	(0.4)
School size	1188.29	(689.1)
Percent 10th graders white	76.44	(29.1)
Percent school graduates who attend four-year college	39.92	(18.6)
Percent teachers with M.A. at school	50.24	(22.5)
Percent school from single-parent families	28.79	(16.2)
Teacher Characteristic Variables		
Female teacher	0.46	(0.5)
Black teacher	0.04	(0.2)
Hispanic teacher	0.02	(0.1)
Asian teacher	0.01	(0.1)
Teacher years experience at secondary level	15.50	(9.0)
Teacher is certified	0.96	(0.2)
Teacher certified in math	0.94	(0.2)
Teacher has B.A.	0.77	(0.4)
Teacher has B.A. major in math	0.52	(0.4)
Teacher has M.A. or more	0.50	(0.5)
Teacher has M.A. major in math	0.17	(0.4)
Teacher Behavior Variables	0.72	(0.4)
Teacher feels very well prepared		
Teacher has no control over discipline	0.04	(0.2)
Teacher has no control over technique	0.01	(0.1)
Teacher has no control over context	0.19	(0.4)
Percent teacher time instructing small groups	11.23	(12.6)
Percent teacher time instructing individuals	18.00	(16.4)
Percent teacher time maintaining order	8.39	(14.9)
Percent teacher time doing administrative tasks	5.52	(6.4)

(continued)

Table 10.1. (*Continued*)

Teacher uses oral questions frequently	0.75	*(0.4)*
Teacher emphasizes problem solving	0.24	*(0.4)*
Class Variables		
Student class size	23.37	(7.0)
Percent minority students in class	20.38	(28.4)
Sample Size	5149	

Results: Basic Models

We begin by estimating standard educational production functions by ordinary least squares. Throughout our analyses, the dependent variable is the tenth-grade mathematics test score. We group our explanatory variables into four sets: *individual and family background variables* (including sex, race/ethnicity, parental education, family structure, family income, and eighth-grade math test score), *school variables* (including urbanicity and regional dummies, school size, the percentage of students at the school who are white, the percentage of students at the school who are from single-parent families, the percentage of students who typically go on to four-year colleges, and the percentage of teachers at the school with at least a Master's degree), *teacher characteristics variables* (including sex, race/ethnicity, years of experience at the secondary level, whether the teacher is certified in mathematics, the teacher's degree level, and whether his or her BA major or MA major are in math), and *classroom variables* (class size and percentage of minority students). As noted previously, a unique feature of our NELS data is that the teacher and classroom variables are specific to the students tenth-grade mathematics class.

Table 10.2 contains the basic model results. We do not show the coefficients of individual and family background variables though they are included in each model. These variables alone explain approximately three-quarters of the variation in the tenth-grade math score. In our sample, most of the estimated coefficients of these variables are statistically significant in the expected direction. For example, years of parental education and family income are positively related to test scores, a result typical of the literature. Black and Hispanic students have lower predicted scores on average, as do those from families with no mother in the household.

School and Teacher Characteristics Variables

Column 1 of Table 10.2 shows the set of estimated coefficients for the school characteristics. Only two of the ten variables are statistically significant, though most coefficients are the expected sign. However, an *F*-test of the hypothesis that the coefficients of the schooling variables are jointly equal to zero is rejected

Table 10.2. Educational Production Functions with Individual, School, Teacher, and Class Variables (absolute value *t*-statistics) Standard OLS Models

	Dependent Variables: 10th-Grade Test Score				
Category	*(1)*	*(2)*	*(3)*	*(4)*	*(5)*
School Variables					
Urban	-0.329 (1.1)	-0.286 (1.0)	-0.346 (1.2)	-0.308 (1.1)	-0.135 (0.5)
Rural	-0.294 (1.3)	-0.326 (1.4)	-0.240 (1.0)	-0.257 (1.1)	-0.307 (1.3)
Northeast	0.583 (1.9)	0.578 (1.9)	0.451 (1.5)	0.450 (0.1)	0.406 (1.3)
North central	-0.011 (0.0)	-0.036 (0.1)	0.047 (0.2)	0.035 (0.1)	-0.053 (0.2)
West	0.059 (0.2)	0.215 (0.7)	0.178 (0.6)	0.303 (1.0)	0.285 (0.9)
School size*	0.167 (0.9)	0.083 (0.4)	0.116 (0.6)	0.057 (0.3)	0.072 (0.4)
Percent white school	-0.006 (1.3)	-0.007 (1.5)	-0.005 (1.2)	-0.006 (1.3)	-0.027 (4.9)
Percent four-year college	0.022 (3.6)	0.019 (3.2)	0.020 (3.4)	0.019 (3.1)	0.019 (3.1)
Percent teachers with M.A.	-0.001 (0.1)	-0.001 (0.3)	-0.002 (0.4)	-0.002 (0.5)	-0.002 (0.5)
Percent single-parent families	-0.004 (0.7)	-0.006 (0.9)	-0.000 (0.0)	-0.002 (0.3)	-0.003 (0.5)
Teacher Characteristic Variables					
Female teacher	—	0.580 (2.9)	—	0.508 (2.5)	0.528 (2.7)
Black teacher	—	-1.178 (2.2)	—	-0.774 (1.5)	-0.588 (1.1)
Hispanic teacher	—	0.716 (1.0)	—	0.795 (1.1)	1.271 (1.8)
Asian teacher	—	0.745 (0.9)	—	1.156 (1.4)	1.382 (1.6)
Teacher years experience at secondary level	—	0.019 (1.6)	—	0.014 (1.2)	0.011 (1.0)
Teacher is certified	—	-2.499 (2.5)	—	-1.780 (1.8)	-1.600 (1.6)
Teacher certified in math	—	2.271 (2.4)	—	-0.801 (2.6)	-0.748 (2.4)
Teacher has B.A.	—	-0.797 (2.6)	—	-0.801 (2.6)	-0.748 (2.4)
Teacher has B.A. major in math	—	0.867 (3.7)	—	0.756 (3.2)	0.717 (3.1)
Teacher has M.A. or more	—	-0.134 (0.5)	—	-0.205 (0.8)	-0.130 (0.5)
Teacher has M.A. major in math	—	0.632 (2.2)	—	0.617 (2.2)	0.590 (2.1)

Table 10.2. (*Continued*)

Teacher Behavior Variables					
Teacher feels very well prepared	—	—	0.896 (4.3)	0.711 (3.3)	0.748 (3.5)
Teacher has no control over discipline	—	—	-0.553 (1.1)	-0.482 (1.0)	-0.403 (0.8)
Teacher has no control over technique	—	—	-3.882 (3.7)	-3.740 (3.5)	-3.532 (3.3)
Teacher has no control over context	—	—	0.350 (1.4)	0.384 (1.5)	0.421 (1.7)
Percent teacher time instructing small groups	—	—	-0.018 (2.8)	-0.015 (1.9)	-0.015 (1.8)
Percent teacher time instructing individuals	—	—	-0.028 (4.5)	-0.026 (4.2)	-0.024 (3.9)
Percent teacher time maintaining order	—	—	-0.020 (3.0)	-0.021 (3.0)	-0.021 (3.1)
Percent teacher time doing administrative tasks	—	—	0.018 (1.2)	0.018 (1.1)	0.015 (0.9)
Teacher uses oral questions frequently	—	—	0.701 (3.1)	0.577 (2.5)	0.601 (2.6)
Teacher emphasizes problem solving	—	—	-0.342 (1.6)	-0.371 (1.7)	-0.377 (1.7)
Class Variables					
Student class size	—	—	—	—	0.031 (2.1)
Percent minority students in class	—	—	—	—	-0.039 (6.3)
Individual and Family Background Variables	Yes	Yes	Yes	Yes	Yes
Adjusted R^2	.763	.766	.768	.770	.772

Note: *Coefficient multiplied by 1,000.

at the 1 percent level. This implies that *as a group* the schooling characteristics do affect student test scores. Column 2 shows the results when teacher characteristics are added to the model. Here eight of the 11 teacher variables have statistically significant coefficients. An *F*-test of the hypothesis that the coefficients of the teacher characteristics are jointly equal to zero is easily rejected at the 1 percent level. This implies that as a group the teacher characteristics do affect student test scores. We find that students of more experienced teachers have higher scores. Female teachers are associated with higher test scores and black teachers with lower test scores; the relationship between teacher race, ethnicity, and gender and student scores is likely more complex than this simple linear model allows but we do not explore the issue here (for a more detailed analysis of this issue Ehrenberg, Goldhaber, & Brewer, 1995).

The teacher characteristic results reported in Table 10.2 contrast with many earlier studies which have found little relationship between such variables and student achievement. As shown in column 2 of Table 10.2, the estimated coefficients of whether the teacher has a Bachelor's degree and whether she has a Master's degree are negative (and in the case of BA, statistically significant) relative to the omitted category of no degree. This implies that teachers with Bachelor's and Master's degrees are less effective than those without any degree, clearly a counterintuitive finding. The results for teacher certification are similar in that we find the coefficient on teacher certification to be statistically significant and negative. However, these models also include dummy variables indicating whether the teacher is certified in math, has a Bachelor's in math, and has a Master's in math. These variables allow us to distinguish between teachers who are teaching mathematics classes and have a major in math (BA or MA), teaching mathematics classes and are certified in math, and those who are teaching mathematics but do not have specific math training. For example, the total effect of a teacher having a Bachelor's degree in math is the sum of the coefficients of BA and BA major in math, whereas the effect of teacher who has a non-math BA is simply the coefficient of the BA variable. We discuss these results further below.

Teacher Behavior Variables

As noted previously, many education researchers have suggested that teacher characteristics do not begin to capture the ways in which teachers affect student achievement. In particular, teacher behavior is typically not included in educational production function analyses which treats the process of schooling as a "black box." Few attempts have been made to utilize such concepts in multivariate analysis, partly because measuring them is extremely difficult (Brewer & Stasz, 1996), and partly because the choice of variables for inclusion is somewhat arbitrary. One useful effort to do this (using data on elementary schools) is Murnane and Phillips (1981). They include, in production functions explaining vocabulary test scores, variables such as the percentage of time the teacher devotes in teaching in subgroups, demonstrations, individualized work, and whether the teacher views his or her job as explaining subject matter. The authors find that these behavior variables explain a larger proportion of test score variance than teacher characteristics alone, although a model including *both* characteristics and behavior measures best fits the data.

Similar information on teacher behavior is included in NELS. This includes the percentage of the teacher's time in class devoted to small groups and individualized instruction, the percentage of time maintaining order and doing administrative tasks, whether the teacher uses oral questions frequently and emphasizes problem solving, whether the teacher has no control over curriculum content, teaching technique and disciplinary policy, and whether the teacher feels well prepared. We selected these items partly to replicate the Murnane and Phillips analysis at the secondary level and partly to capture newer conceptions of mathematics teaching (small groups, problem solving) as reflected in the National Council of Teachers of Mathematics (NCTM, 1991) standards.

Column 3 of Table 10.2 shows a model including teacher behavior but not teacher characteristics; column 4 includes both sets of variables. The model with teacher behavior variables is marginally better at explaining test scores than the characteristics model, but including both sets of variables together provides for the best fit of the data, as was the case in Murnane and Phillips (this is confirmed by *F*-tests of the hypotheses that the added coefficients are jointly equal to zero). However, the estimated coefficients of the teacher characteristics variables do not change much when the behavior variables are added; the reverse is also true. This suggests that there may be no strong relationship between our set of teacher characteristics and any teacher behavior variable. We confirmed this in auxiliary regressions of teacher behavior on teacher characteristics.[10]

Several results are worth highlighting. Students with teachers who have little or no control over their teaching technique have significantly lower test scores. Variables associated with NCTM-type standards, such as teaching in smaller groups and emphasizing problem solving, appear to lower student math scores. This does not necessarily imply that these teaching techniques are invalid but may simply indicate that these methods do not improve achievement on traditional standardized tests. (It is also possible that these methods improve scores for some individuals or subgroups of students but not for all students on average.) One caution is that the NELS teacher behavior items are based on teacher survey responses (self-reports). Burstein, McDonnell, Van Winkle, Ormseth, Mirocha, and Guitton (1995) assess the validity of NELS instructional strategy items using an enhanced version of the teacher survey instruments, teacher interviews, and collection of class assignments and homework with a small sample of teachers in the original NELS sample. Their results suggest such survey items and response scales need to be treated cautiously.

Having said this, it is interesting to note that Murnane and Phillips (1981) report similar findings; for example, they found a similar statistically significant negative coefficient on percentage of time devoted to subgroup instruction for sixth-grade vocabulary. All of our results, at the secondary level, are strikingly consistent with their results at the primary level. On net, it appears that these types of variables should be included in educational production functions but since our focus here is primarily on biases in traditional education production functions, we do not discuss these variables and do not include them in subsequent models shown. Our conclusions below are unaffected by this simplification.

Classroom Variables

Column 5 of Table 10.2 adds two classroom-level variables to the production function: class size, which is positively associated with achievement, and the percentage of minority students in the class, which is negatively associated with achievement. The class size result is counterintuitive (though the magnitude of the estimated coefficient is small) but not atypical of production function results. As noted earlier, Akerhielm (1995) found a similar result with NELS data which she attributed to the nonrandom assignment of students to classes. Again an *F*-test comparing models easily rejects the hypothesis that the coefficients of these classroom variables are jointly equal to zero, suggesting that they should be included in our model.

Further Models

Several extensions of the basic analyses reported earlier were conducted in order to determine the robustness of our results. We pursued two lines of inquiry. First, to what extent are our results affected by the use of the rich, microlevel information in NELS? Second, can we be reasonably sure that our results are not biased by the omission of important explanatory factors from our model of student achievement? We provide some evidence here on these issues. Readers are referred to Goldhaber and Brewer (forthcoming) for details of the econometric methods used and a complete set of results.

The importance of data specificity can be seen most explicitly by examining our results with regard to the estimated effect of teacher educational qualifications on student achievement. Traditional education production functions do not include subject-specific teacher degree and certification information. However, at least in our sample, the use of this more detailed data is critical in interpreting the effects of these teacher characteristics on student achievement. This can be seen in Table 10.3.

Table 10.3. Comparison of Selected Coefficients from Educational Production Functions (absolute value *t*-statistics)

	(1)	*(2)*	*(3)*	*(4)*
Teacher years experience at secondary level	0.022 (1.9)	0.023 (2.0)	0.016 (1.4)	0.017 (1.5)
Teacher is certified	-0.072 (0.1)	-0.305 (0.5)	-2.265 (2.3)	-2.412 (2.4)
Teacher certified in math	—	—	2.150 (2.2)	2.024 (2.1)
Teacher has B.A.	-0.183 (0.7)	-0.047 (0.2)	-0.736 (2.4)	-0.578 (1.8)
Teacher has B.A. major in math	—	—	0.823 (3.5)	0.748 (3.1)
Teacher has M.A. or more	0.130 (0.5)	0.162 (0.7)	-0.055 (0.2)	-0.010 (0.0)
Teacher has M.A. major in math	—	—	0.606 (2.2)	0.536 (1.9)
State Dummies	No	Yes	No	Yes
Adjusted R^2	.767	.770	.768	.772

Notes: All models contain individual and family background variables, school variables, teacher characteristic variables (except as specified in table), and class variables. Table 10.1, column (3), reproduced in Table 10.2, column (3).

Column 1 of Table 10.3 shows a model in which math-specific degree and certification information is omitted. This result would lead one to the conclusion that teacher degrees and certification have no impact on student achievement. This is in line with much of the previous literature. However, the results reported in column 3 of Table 10.3 would lead to a quite different conclusion. Here we show that a teacher with a BA in math or an MA in math has a statistically significant *positive* impact on students' achievement, while a teacher with a non-math BA or an MA has a negative impact on students' math achievement (all relative to the omitted category of no degree). We find similar results with teacher certification when we estimate models with (columns 1 and 3 of Table 10.3) and without (columns 2 and 4 of Table 10.3) a set of state dummy variables indicating the location of the high school. We estimate models with state dummies to control for the differing certification requirements across states.

The second concern about our results is the possibility that they are biased due to the presence of unobservable determinants of student achievement that are systematically correlated with the variables we do include. The NELS data permit us to estimate a variety of econometric models that shed some light on this issue (including one-way fixed and random effects models). Our results suggest that while observable school, teacher, or class variables account for a relatively small fraction of the variation in student test scores, unobservable factors associated with these schooling characteristics are important. For example, regression models that include teacher "fixed effects" (that capture both unobservable and observable factors associated with teachers and schools in our sample) can explain almost 89 percent of the variation in test scores, as compared to under 80 percent for models which only include observable variables (such as those reported in Table 10.2). However, we find that unobservable school, teacher, and class effects do not appear to be correlated with observable variables included in our models, and thus do not cause biased estimates of the effects of these variables in our sample.[11]

CONCLUSION

In this chapter we assess the importance of educational resources in explaining student achievement on a tenth-grade standardized mathematics test. Using data from the *National Educational Longitudinal Study of 1988*, which permit us to link students with a particular teacher and math class, we find that traditional (OLS) educational production function models do show educational resources to be significant in influencing tenth-grade mathematics test scores. Although school-level variables do not, in general, seem to have much affect on student achievement, particular teacher characteristics do. For example, teachers with specific training in mathematics—for example, who are certified in mathematics, and those with Bachelor's or Master's degrees in math—are identified with higher student test scores. Our extension of the work of Murnane and Phillips (1981) also shows that some aspects of teacher behavior influence student achievement, independent of teacher characteristics.

Our findings reported in this chapter regarding teacher traits may be due to the unique nature of the NELS data among national surveys in providing a direct student-teacher-class link. This link enables us to avoid problems with aggregation that may have plagued earlier studies. Also, subtle differences in model specification can result in very different interpretation of whether teachers, for example, affect student outcomes. Student achievement models estimated with variables specifying whether a teacher has a BA or an MA but without the subject of the degrees, show teacher degree to be statistically insignificant; this is also true of teacher certification. When degree subject is added, teachers with mathematics degrees are clearly associated with higher student scores on the NELS mathematics test.

ACKNOWLEDGMENT

Brewer thanks RAND for funding to support this research.

NOTES

1. The pessimistic view that America's public schools are failing and need dramatic reform is not universally shared; see, for example, Bracey (1991), Grissmer, Kirby, Berends, and Williamson (1994), and Berliner and Biddle (1995). A recent report challenging the conventional wisdom about the extent to which spending in schools has increased is Rothstein (1995).

2. The relationship between wages and test scores has recently been examined by Murnane, Willett, and Levy (1995).

3. The focus of our discussion is on schooling inputs. It is quite likely, however, that there are unobservable student characteristics such as motivation, which are typically omitted from educational production functions.

4. There is some evidence that public schools and school districts are in fact inefficient with regard to resource allocation (Callan & Santerre, 1990) and production technology (Grosskopf, Hayes, Taylor, & Weber, 1995).

5. With a few exceptions most educational production function studies use simple linear ordinary least squares (OLS) regression. However, schooling inputs may not be exogenous as OLS requires. For example, highly motivated teachers likely locate in a nonrandom manner across school districts, schools within a district, and classes within a school. Parents similarly self-select their children into school districts that most closely match their preferences, including willingness to pay for schooling (Goldhaber, 1996). There have been few attempts to control for this potential endogeneity (Ehrenberg & Brewer 1994, 1995; Akerhielm 1995) largely because it is extremely difficult to find suitable instruments for schooling inputs, so it is difficult to assess its importance.

6. For a discussion of aggregation bias in this context see Hanushek, Rivkin, and Taylor (forthcoming).

7. The tenth-grade test was administered during the spring of tenth grade. Unfortunately no data are available on students' ninth-grade schooling characteristics.

8. IRT is a method that uses the pattern of right, wrong, and omitted responses to the questions actually administered on each test and the difficulty, discriminating ability, and "guessability" of each question to place each student on a continuous scale regardless of the test he or she was given (see Rock & Pollock, 1991, for a discussion of IRT and the NELS data).

9. A recent study of the effects of tracking on student achievement using the NELS data is Argys, Rees, and Brewer (1996).

10. We ran a series of simple regression models in which each teacher behavior variable was regressed against observable teacher characteristics variables. There were few consistent results across models. Further, the fit of these models was extremely poor: the R^2 was typically less than 5 percent.

11. This result is obtained via the estimation of random effects models and performing statistical tests that permit us to compare the estimates of our various models.

REFERENCES

Akerhielm, K. (1995). Does class size matter? *Economics of Education Review, 14* (3), 229-241.

Argys, L., Rees, D.I., & Brewer, D.J. (1996, Fall). The impact of ability grouping on high school achievement: Evidence from NELS. *Journal of Policy Analysis and Management, 15* (3).

Berliner, D.C., & Biddle, B.J. (1995). *The manufactured crisis: Myths, fraud and the attack on America's public schools.* Reading, MA: Addison-Wesley.

Betts, J. (1995). Does school quality matter: Evidence from the national longitudinal survey of youth. *Review of Economics and Statistics, LXXVII* (2), 231-250.

Boardman, A.E., & Murnane, R.J. (1979). Using panel data to improve estimates of the determinants of educational achievement. *Sociology of Education, 52*, 113-121.

Boyd, W.L., & Hartman, W.T. (1988). The politics of educational productivity. In D.H. Monk & J. Underwood (Eds.), *Microlevel school finance.* Cambridge: Ballinger.

Bracey, G.W. (1991). *Phi Delta Kappan, 73* (2), 104-117.

Brewer, D.J. (1996). Does more school district administration lower educational productivity? Some evidence on the "Administrative Blob" in New York public schools. *Economics of Education Review, 15* (3), 111-124.

Brewer, D.J., & Stasz, C. (1996). Enhancing opportunity to learn measures in NCES data. In *From data to information: New directions for the national center for education statistics.* Washington, DC: National Center for Education Statistics.

Brophy, J., & Good, T. (1986). Teacher behavior and student achievement. In M. Whittrock (Ed.), *Handbook of research on teaching.* New York: MacMillan.

Burstein, L., McDonnell, L.M., Van Winkle, J., Ormseth, T., Mirocha, J. & Guitton, G. (1995). *Validating national curriculum indicators.* Santa Monica, CA: RAND. (DRU-1086-NSF.)

Callan, S.J., & Santerre, R.E. (1990). The production characteristics of local public education: A multiple product and input analysis. *Southern Economic Journal, 57* (2), 468-480.

Card, D., & Krueger, A. (1992). Does school quality matter? Returns to education and the characteristics of public schools in the United States. *Journal of Political Economy, 100* (1), 1-40.

Coleman, J.S. et al. (1966). *Equality of educational opportunity.* Washington, DC: Department of Health, Education, and Welfare.

Ehrenberg, R.G., & Brewer, D.J. (1994). Do school and teacher characteristics matter? Evidence from high school and beyond. *Economics of Education Review, 13* (1), 1-17.

________ . (1995). Did teachers' verbal ability and race matter in the 1960s? Coleman revisited. *Economics of Education Review, 14* (1), 1-23.

Ehrenberg, R.G., Goldhaber, D.D., & Brewer, D.J. (1995). Do teachers' race, gender, and ethnicity matter? Evidence from NELS88. *Industrial and Labor Relations Review, 48* (3), 547-561.

Evans, W.N., & Schwab, R.M. (1995, November). Finishing high school and starting college: Do Catholic schools make a difference. *Quarterly Journal of Economics.*

Ferguson, R. (1991). Paying for public education: New evidence on how and why money matters. *Harvard Journal on Legislation, 28*, 465-498.

Goldhaber, D.D. (1996). Public and private high schools: Is school choice an answer to the productivity problem? *Economics of Education Review, 15* (3).

Goldhaber, D.D., & Brewer, D.J. (forthcoming). Why don't schools and teachers seem to matter? Assessing the impact of unobservables on educational productivity. *Journal of Human Resources.*

Good., T.L. (1994). Teaching effects and teacher evaluation. Draft manuscript of a chapter to appear in the second edition of the *Handbook of Research on Teacher Education.*

Grissmer, D., Kirby, S., Berends, M., & Williamson, S. (1994). *Student achievement and the changing American family.* Santa Monica, CA: RAND.

Grosskopf, S., Hayes, K., Taylor, L., & Weber. W. (1995). On competition and school efficiency. Paper presented to Western Economic Association Conference, July.

Hanushek, E.A. (1979). Conceptual and empirical issues in the estimation of education production functions. *Journal of Human Resources, 14* (3), 351-388.

________ . (1986). The economics of schooling: Production and efficiency in the public schools. *Journal of Economic Literature, XXIV* (3), 1141-1178.

________ . (1989). The impact of differential expenditures on school performance. *Educational Researcher, 18* (4).

________ . (1994). Reply to Hedges et al. *Educational Researcher, 23* (4).

Hanushek, E.A., Rivkin, S., & Taylor, L. (forthcoming). Aggregation and the estimated effects of school resources. *Review of Economics and Statistics.*

Hedges, L., Laine, R. & Greenwald, R. (1994). A meta analysis of the effects if differential school inputs on student outcomes. *Educational Researcher, 23* (3), 5-14.

Laine, R.D., Greenwald, R., & Hedges, L. (1995). Money does matter: A research synthesis of a new universe of education production function studies. In L.O. Picus & J.L. Wattenbarger (Eds.), *Where does the money go? Resource allocation in elementary and secondary schools.* Thousand Oaks, CA: Corwin Press.

Monk, D.H. (1992). Educational productivity research: An update and assessment of its role in education finance reform. *Educational Evaluation and Policy Analysis, 14*, 307-332.

Monk, D.H., & King, J. (1994). Multi-level teacher resource effects on pupil performance in secondary mathematics and science: The role of teacher subject matter preparation. In R.G. Ehrenberg (Ed.), *Contemporary policy issues: Choices and consequences in education*. Ithaca, NY: ILR Press.

Murnane, R.J., & Phillips, B. (1981). What do effective teachers of inner-city children have in common? *Social Science Research, 10*, 83-100.

Murnane, R.J., Willett, J.B., & Levy, F. (1995). The growing importance of cognitive skills in wage determination. *Review of Economics and Statistics, LXXVII* (2), 251-266.

National Center for Education Statistics (NCES). (1994). *Digest of educational statistics*. Washington, DC: U.S. Department of Education, NCES 94-115.

National Council of Teachers of Mathematics (NCTM). (1991). *Professional standards for teaching mathematics*. Reston, VA: NCTM.

Picus, L.O., & Wattenbarger, J.L. (Eds.). (1995). *Where does the money go? Resource allocation in elementary and secondary schools*. Thousand Oaks, CA: Corwin.

Purkey, S.C., & Smith, M.S. (1983). Effective schools: A review. *The Elementary School Journal, 83* (4), 427-452.

Rock, D.A., & Pollock, J.M. (1991). *Psychometric report for NELS:88 base year test battery*. Washington, DC: National Center for Education Statistics.

Rosenholtz, S.J. (1985). Effective schools: Interpreting the evidence. *American Journal of Education, 93* (3), 352-388.

Rothstein, R. (1995). *Where's the money gone? Changes in the level and composition of education spending*. Washington, DC: The Economic Policy Institute.

Chapter 11

THE DEMOGRAPHICS OF SCHOOL FUNDING VARIATIONS:

AN ATTEMPT TO INFORM THE DEBATE

Linda Hertert

It is not uncommon to read that concerns over efficiency have lately eclipsed concerns over equity in school finance debates. The implication is that the one, say efficiency, can only be considered at the expense of the other, equity. This establishes a false dichotomy that does little to advance attempts to restructure the distribution of fiscal resources in support of increased student performance. Such statements also suggest that the most complicated aspect of restructuring educational resources is deciding which conceptual framework to employ—efficiency or equity. The task is far from this simple.

As Hanushek (1996) writes, "... equity and efficiency are inextricably linked" (p. 33). It is with this understanding that this research is presented. To argue simply for equity is to ignore how fiscal resources are used and for

Advances in Educational Productivity, Volume 6, pages 265-279.

ISBN: 0-7623-0082-5

what purpose. Further, simply to argue for equity is to ignore the issue of adequacy. It is possible for resources to be distributed equitably and be completely inadequate to the intended purpose.

Informed discussions about these complex issues should rest on some factual base. It is the intent of this paper to contribute to that base by exploring the connections between the distribution of educational resources and certain demographic characteristics of the recipient pupils, districts, and communities. It is not the intent to suggest that educational outcomes would improve if the variations in fiscal resources districts receive were reduced. That debate is left to others.

METHODOLOGY

This study examines the relationship between school district revenues and certain demographic measures at one point in time: 1989-1990.

Data

The data are taken from two sources: (1) the fiscal information is from the Common Core of Data (CCD) compiled by the National Center for Education Statistics (NCES), and (2) the demographic indicators are from the 1990 Census School District Special Tabulation jointly compiled by NCES and the United States Bureau of the Census.[1]

The population is defined as all "regular" operating school districts in the United States, omitting those devoted specifically to the education of special pupil populations. Merging the two data sources results in a database of 14,528 school districts with a combined student enrollment of 39,542,319 pupils.

Variables

Per pupil revenues are computed as the sum of all operating revenues received by a school district for FY1990, excluding federal funds and monies raised for capital improvements and debt service. Where noted, revenues are adjusted for district-type and interstate cost differentials (for specifics on adjustment computation, see Hertert, Busch, & Odden, 1994). District and community characteristics are described as follows.

Districts

Type: elementary, high school, or unified.
Size: school district enrollment.
Ethnicity: percent of nonwhite pupils attending school in the district.

Location: five categories including three inside metropolitan areas (central city, large urban areas with greater than 2,500 residents, and small urban areas with fewer than 2,500 residents) and two outside metropolitan areas (suburban areas and rural areas).

Community

Poverty: percent of children ages 0-19 in the district who are living below the 1989 poverty level.
Per capita income: 1989-1990 per capita income of district residents.
Location: eight census regions: New England, Mideast, Southeast, Great Lakes, Plains, Southwest, Rockies, and Far West.[2]

Analyses

The relationship between district revenues and demographic characteristics is assessed in two ways. First, descriptive statistics (the mean and coefficient of variation) are computed for per pupil revenues by pupil quartiles, by each characteristic, and by nested characteristics (i.e., revenues by pupil quartile by district location). Next, a stepwise multiple regression is computed to determine the contribution of the characteristics to the variation in per pupil, unadjusted revenues. Dummy coding is used for district type, district location, and community type with unified districts, rural districts, with the Plains region used as reference groups. Interactions are also computed.

The results of these analyses are then used to adjust the distribution of educational revenues. Equity statistics (Berne & Stiefel, 1984) are computed to measure the effects of these adjustments. The next section reports these findings.

FINDINGS: THE DEMOGRAPHICS OF EDUCATIONAL FUNDING PATTERNS

Across the country, 1989-1990 per pupil revenues varied substantially (COV equals 35.1%) around a mean of $4,713. Adjusting revenues does not dramatically affect these levels: $4,687 at 28.3 percent. Tables 11.1 and 11.2 describe the distribution of these revenues by various categories.

Revenues by Pupil Quartiles

Table 11.1 shows the weighted mean and coefficient of variation for adjusted and unadjusted per pupil revenues measured by pupil quartile (allowed to break by district).

Table 11.1. Nationwide Per Pupil Revenues by Pupil Quartiles, 1989-1990

			Unadjusted		*Adjusted*	
	Pupils (#)	*Districts* (#)	*Mean** ($)	*CV** (%)	*Mean** ($)	*CV** (%)
Quartile one (lowest)	9,880,064	3,824	3,121	12.91	3,324	8.92
Quartile two	9,889,959	3,447	4,066	5.62	4,101	5.46
Quartile three	9,518,750	6,051	4,852	5.57	4,866	5.40
Quartile four (highest)	10,253,546	4,206	6,742	21.78	6,453	0.00

Note: *Pupil-weighted.

The range of average, unadjusted revenues is more than $3,600 per pupil across the quartiles. Districts in the second and third quartile receive roughly the same amount per pupil (COV equals 5.62% and 5.57%, respectively) while those districts in the highest and lowest quartiles do not. Still, the majority of pupils in the fourth quartile receive on average more than the majority of pupils in any other quartile (one standard deviation equals +/-$1,468 in the highest quartile).

The range across quartiles is reduced by $500 per pupil when revenues are adjusted for district-type and interstate cost differences. The lowest and highest quartiles, however, continue to show the greatest variation in per pupil revenues across districts.

Revenues by Demographic Indicators

Table 11.2 lists the average per pupil revenues for various district and community characteristics.

District Type

High school districts received on average more money per pupil than either elementary or unified districts. There is substantial variation in per pupil revenues across high school districts (COV equals 32.89%), though not as much as measured across elementary districts (38.15%). When dollars are adjusted (including an adjustment for district type), elementary districts show the highest average per pupil revenues and high school districts the lowest. Revenue differences remain large within each district type (from 27.2 to 39.26%).

District Size

The distribution of per pupil revenues by district size is mixed. Smaller districts (up to 500 pupils) receive on average more dollars, both unadjusted and adjusted, than larger districts. Economies of scale appear to be operating,

Table 11.2. Nationwide Per Pupil Revenues by Various Demographic Indicators, 1989-1990

	Pupils (#)	Districts (#)	Unadjusted Mean* ($)	Unadjusted CV* (%)	Adjusted Mean* ($)	Adjusted CV* (%)
District type						
Elementary districts	2,193,172	3,240	4,683	38.15	4,932	39.26
High school districts	817,657	541	6,340	32.89	4,668	36.20
Unified districts	36,531,490	10,747	4,678	32.36	4,673	27.20
District size						
0-500 pupils	1,113,300	5,177	5,164	43.74	5,526	39.71
500-1,000 pupils	1,719,770	2,386	4,751	37.30	4,848	32.61
1,000-1,500 pupils	1,933,336	1,566	4,687	37.12	4,654	31.18
1,500-5,000 pupils	10,343,059	3,827	4,712	39.05	4,661	32.09
5,000-10,000 pupils	6,394,658	924	4,655	35.74	4,609	29.05
over 10,000	18,038,196	648	4,706	25.81	4,668	22.86
Pupil ethnicity						
0-10% minority	15,210,038	10,090	4,720	34.41	4,742	29.54
11-25% minority	7,976,388	2,088	4,707	35.40	4,675	29.82
26-50% minority	8,899,041	1,492	4,534	31.01	4,570	27.46
51-75% minority	5,555,010	549	5,007	28.70	4,778	24.43
over 75%	1,901,842	309	4,665	32.65	4,598	25.34
District location						
Central city	12,100,717	641	4,757	26.30	4,677	22.76
Large urban	12,678,903	2,560	5,212	34.38	5,036	30.46
Small urban	4,715,387	2,634	4,719	33.43	4,701	28.69
Suburban	3,425,465	1,409	3,940	27.09	4,158	23.17
Rural	6,621,847	7,284	4,071	34.74	4,305	29.88
Children in poverty						
0-10%	11,613,105	4,875	5,476	34.32	5,291	30.14
11-25%	17,258,794	6,509	4,361	27.28	4,450	24.58
26-50%	10,235,186	2,750	4,501	31.42	4,452	25.18
51-75%	428,840	328	3,276	27.51	3,536	24.19
over 75%	6,394	66	4,333	41.58	4,638	40.73
Per capita income						
Pupil quartile one	9,884,677	6,419	3,737	29.74	3,996	24.92
Pupil quartile two	9,885,387	3,373	4,246	25.77	4,436	22.92
Pupil quartile three	9,885,690	2,325	4,492	24.77	4,699	22.81
Pupil quartile four	9,886,565	2,411	4,935	30.91	5,619	27.97
Community location						
New England	1,833,958	1,138	6,118	23.95	5,994	24.56
Mideast	6,014,412	1,778	6,683	25.69	6,090	23.12
Southeast	9,746,013	1,648	3,969	28.64	4,386	24.62
Great Lakes	6,860,652	2,845	4,699	25.81	4,600	24.87
Plains	2,956,331	2,902	4,497	23.32	4,922	22.32
Southwest	4,847,155	1,926	3,653	18.99	3,817	18.34
Rockies	1,444,509	867	3,913	31.15	4,102	26.93
Far West	5,839,289	1,424	4,690	20.82	4,189	20.02

Note: * Pupil-weighted.

though not absolutely. The largest districts (over 10,000) receive more per pupil ($4,687) than districts with only 1,000 to 1,500 pupils and almost as much as districts with 1,500 to 5,000 pupils ($4,712). When dollars are adjusted, the largest districts receive more money per pupil than all but the smallest districts (less than 1,000 pupils). The differences in the amount of money districts receive within each of these categories are large for both adjusted and unadjusted dollars, ranging from 22.86 to 43.74 percent.

Pupil Ethnicity

The relationship between pupil ethnicity and revenues is also mixed. Differences in per pupil revenues (unadjusted and adjusted) across districts with similar enrollment mixes vary by more than 25 percent around the average for the majority of pupils. Unadjusted revenues are highest in districts where 51 to 75 percent of the enrollment is nonwhite and lowest in districts with ethnic enrollments of between 26 and 50 percent. Districts with up to 25 percent or over 75 percent receive on average roughly the same amount of money per pupil. These nonlinear relationships remain when calculated with adjusted dollars.

District Location

Districts located in large urban areas receive on average from $455 to $1,272 per pupil more than districts located in any other type of community setting. School districts in suburban areas receive the lowest amount per pupil on average. The ranking of district revenues by location remains the same when adjusted dollars are used. There is, however, substantial variation in the amount of unadjusted and adjusted dollars similarly located districts receive per pupil (COV ranges from 22.76 to 34.38%).

Children in Poverty

The relationship between the percent of school-age children living in poverty and a district's per pupil revenues is not linear. Districts in communities with less than 10 percent of their children living in poverty receive on average the largest amount of money per pupil, both in adjusted and unadjusted dollars. The next highest per pupil, unadjusted revenues, however, go to districts where the percent of poverty ranges between 26 to 50 percent. The difference in funding between the highest and the second highest revenue districts is a substantial $975 per pupil.

When adjusted dollars are computed the districts with the second highest per pupil revenues are in communities where more than 75 percent of the school-age children are living at or below the poverty line. Again, the difference in per pupil revenues is substantial between the highest and next highest categories, measuring $653 per pupil. Variations in average per pupil revenues are large in all cases, ranging from a coefficient of variation of 24.19 to 41.8 percent.

Per Capita Income

There is a direct relationship between per capita income and a district's per pupil revenues, adjusted or unadjusted: the higher the per capita income the higher the per pupil revenues. There is a striking difference in revenue amounts across pupil quartiles. For unadjusted dollars, the top quartile receives on average almost $1,200 per pupil more than the lowest quartile. With adjusted dollars, the difference is over $1,600 per pupil. Variations around the average are large, however, within quartiles, ranging from a coefficient of variation of 22.81 to 30.91 percent.

Community Location

Districts in the Mideast receive on average more dollars per pupil than districts in other regions, and districts in the Southwest receive less. Regional differences in per pupil revenues are reduced when using adjusted dollars, but regional ranking is unaffected. There is less variation in per pupil revenues across districts in the Southwest, measuring a hefty 18.34 percent coefficient of variation for adjusted dollars. The greatest variability is found across districts in the Rockies with a coefficient of variation of 31.15 percent for adjusted revenues.

Characteristics of Per Pupil Revenue Quartiles

Table 11.3 describes the demographic characteristics associated with the distribution of unadjusted revenues as sorted by pupil quartiles and disaggregated by district location. Data using adjusted dollars are not shown here as this recalculation had little effect on the results.

District Location

The figures in the first column of Table 11.3 show that pupils in suburban and rural districts are disproportionately represented in the lowest quartile. Almost half of the children attending schools in suburban areas (46.63%) are in districts in the lowest quartile. Only 8.97 percent of these suburban pupils are enrolled in districts in the top quartile. Rural districts show a similar configuration with 42.17 percent of their pupils attending districts falling into the lowest quartile and only 11.76 percent of their pupils in the highest quartile. These districts, suburban and rural, receive on average less than the other districts in the bottom quartile ($3,105 and $2,978 per pupil, respectively).

Table 11.3. Nationwide School Characteristics by Unadjusted Per Pupil Revenues and District Locations, 1989-1990

	Pupils (%)	*Districts (%)*	*Revenues ($)*		*Size (#)*		*Minority Pupils (%)*		*Children in poverty (%)*		*Per capita income ($)*	
			*Mean**	*CV**	*Mean**	*CV**	*Mean**	*CV**	*Mean**	*CV**	*Mean**	*CV**
Quartile One (Lowest)												
Central city	21.96	25.12	3,272	9.48	16,508	138.78	39.44	57.63	27.59	39.76	12,172	19.62
Large urban	13.94	14.34	3,147	11.94	4,814	169.45	18.02	95.01	15.64	71.80	12,978	33.02
Small urban	22.61	20.46	3,105	13.20	1,978	156.88	11.51	119.55	16.78	51.01	11,706	21.60
Suburban	46.63	43.93	3,105	11.89	2,580	80.47	23.99	90.99	25.38	46.89	10,602	16.63
Rural	42.17	29.35	2,978	15.32	1,306	139.13	19.67	123.95	26.92	44.91	9,467	17.91
Quartile Two												
Central city	22.14	28.55	4,068	5.32	14,641	118.47	35.21	62.03	20.84	45.15	13,814	19.55
Large urban	25.34	20.74	4,114	5.50	6,051	1652.56	24.08	79.98	12.94	61.13	15,571	26.15
Small urban	27.26	24.98	4,023	5.70	1,954	163.99	11.45	122.88	12.31	66.21	13,124	19.67
Suburban	26.86	26.96	4,020	5.72	2,447	87.39	17.12	109.35	20.50	42.20	11,616	15.11
Rural	27.06	23.33	4,032	5.75	1,055	165.34	17.15	128.28	19.57	51.66	10,900	312.48
Quartile Three												
Central city	26.04	24.34	4,830	5.75	20,198	193.33	43.90	60.00	24.16	43.58	13,729	19.03
Large urban	26.07	21.21	4,874	5.48	6,088	168.28	19.28	91.23	10.74	68.99	16,983	30.66
Small urban	25.48	21.72	4,872	5.37	2,101	144.23	6.53	139.97	10.02	65.27	14,193	20.43
Suburban	17.54	18.38	4,827	5.34	2,320	96.88	9.88	137.55	15.97	47.09	12,369	16.07
Rural	19.02	20.88	4,839	5.54	828	161.37	10.98	160.56	17.65	52.10	11,336	52.69
Quartile Four (Highest)												
Central city	29.86	22.00	6,298	13.44	25,622	364.58	57.30	36.44	27.72	30.34	15,934	24.78
Large urban	34.65	43.71	7,101	23.85	3,926	311.20	20.60	92.67	8.61	81.77	19,972	32.41
Small urban	24.64	32.84	6,813	23.11	1,343	127.27	5.45	125.69	6.24	78.53	18,516	35.78
Suburban	8.97	11.00	6,312	15.00	1,983	100.17	9.92	102.82	13.90	45.97	13,305	19.37
Rural	11.76	26.44	6,843	25.74	404	169.91	8.54	201.41	15.26	62.58	12,679	26.53

Note: *Pupil-weighted.

The percent of pupils attending school in central cities and small urban districts is roughly the same across the revenue quartiles. The distribution of pupils in large urban districts favors the upper quartiles—only 13.94 percent of these children are in districts in the bottom quartile—and these districts receive on average more money per pupil than others in the same quartile.

District Size

On average, districts with the largest enrollments are found in central cities, and the larger these districts the more likely they are to be in the highest revenue quartile. There are, however, huge variations in average size in each pupil quartile and for each community location. Rural districts have the smallest average size per quartile with size inversely related to revenues. This suggests that the presence of rural districts in the highest quartile may be simply a function of spreading basic, necessary operating costs over very few pupils—most districts in the country with 25 or fewer pupils are located in rural areas.

Pupil Ethnicity

Central city districts have the highest percent of minority pupils in every quartile. The largest average percent of minority pupils (57.3%) is found in central city districts in the highest revenue quartile. The percent of minority pupils in large urban districts is spread fairly evenly across the four quartiles. There is, however, an inverse relationship between the percent of minority pupils enrolled and per pupil revenues received in small urban, suburban, and rural districts—the higher the revenues the smaller the percent of minority pupils.

Children in Poverty

Central city districts in the highest quartile have almost the same percent of children living in poverty (27.72%) as do central city districts in the lowest revenue quartile (27.59%). For other community locations, though, poverty percentages decrease as revenues increase.

Poverty is not exclusively an inner-city phenomenon. Suburban and rural districts in the two lowest revenue quartiles have approximately the same percent of their children living in poverty as do central city districts (25.38, 26.92% and 20.5, 19.57%, respectively). This pattern disappears in the upper half of the distribution with central city districts having 5 to 10 percent more children in poverty on average than either suburban or rural districts. Interestingly, large and small urban districts measure the lowest percentage of poor children in all four quartiles.

Per Capita Income

Per capita income generally increases for all district locations as per pupil revenues increase. The only exception is in central city districts where there is a drop of $85 in average per capita income between the second and third quartiles followed by a rise of $2,205 in the fourth or highest quartile. Rural school districts consistently record the lowest per capita income of any district configuration considered, ranging from a low of $9,467 in the bottom quartile to a high of $12,679 in the highest quartile.

Summary

This section described the distribution of educational funds with respect to various demographic characteristics. The most consistent relationship noted was the positive one between average per capita income and educational revenues. Other comparisons between revenues and district and/or community factors were mixed but still provided some interesting information. For example:

1. Districts enrolling high percentages of minority pupils are not necessarily districts receiving the least amount of revenue per pupil—central city districts in the top revenue quartile had the highest average percent of minority pupils of any district location in any revenue quartile.
2. The majority of school children in central cities are not enrolled in the poor districts (those in the bottom quartile). This is, however, true for children living in suburban and rural areas—on average, large urban districts receive the highest per pupil revenues while suburban and rural districts receive the lowest.
3. Poverty is not exclusive to inner-cities—suburban and rural districts in the bottom revenue quartile had approximately the same percentage of children living below the poverty line as did central city districts.

The next section examines the degree to which differences in per pupil revenues are attributable to the various characteristics discussed previously.

ESTIMATING THE RELATIONSHIP

A stepwise multiple regression is used to estimate the relationship between the demographic variables and the variation in district revenues per pupil. As shown in Table 11.4, the model explains 41.85 percent of the variation across school districts in the country (R^2(adj) = 0.4185, $F = 189.82, p < .000$). Because of the differences in the unit of measure across the variables, this discussion will focus on the relative importance of each factor in terms of Beta.

Table 11.4. Significant Parameter Estimates for Regression Model of Public School Funding Revenue, 1989-1990

	Coefficient	*Beta*
Variables		
Per-capita income	0.593	1.356
Percent minority pupils	-67.960	-0.547
Percent pupils in poverty	96.266	0.562
Interactions		
Minority by:		
Central city	-54.556	-0.184
Large urban	-81.490	-0.350
Small urban	-44.701	-0.137
Suburban	-28.198	-0.093
Poverty by:		
Large urban	54.494	0.139
Small urban	32.056	0.100

Notes: R^2 (adj) = 0.4185
F = 189.82

As indicated in Table 11.4, per capita income is most strongly associated with the differences in educational revenues (Beta = 1.356, $p < .000$). The percent of minority pupils enrolled in a district is inversely related to educational revenues (Beta = -0.547, $p < .000$). A surprisingly strong finding given the observations made earlier. However, the influence of minority status is significantly different across district locations (see interactions), and as shown in Table 11.3 the only tendency toward a positive relationship between minority status and revenues exists in central city locations.

The percent of poor children in a district's community is positively associated with educational revenues (Beta = 0.562, $p < .000$). This variable also has a differential influence across district locations, with significance measured in large ($p < .000$, with Beta = 0.139) and small urban areas ($p < .000$ with Beta = 0.1). The influence of these two factors, minority status and poverty level, is secondary, however, to the association between per capita income and educational revenues.

The analysis indicates that a district's revenues are linked to local capacity as measured by per capita income. This seems reasonable given that (1) a large portion of educational revenues within states is raised locally via taxes, (2) the ability and willingness of taxpayers to support local levies are in part a function of their personal wealth, and (3) much of the revenue raised at the local level is not fully equalized by states. It also seems reasonable that the relative degree of poverty, which is reflected in any calculation of a community's per capita income, should likewise be associated with a district's revenues.

The analysis also indicates an inverse relationship between minority status and revenues. Minority pupils are concentrated in central city districts—districts in which large numbers of pupils often qualify for supplemental funding. The data used here excluded federal revenues, but included supplemental funding raised within states to support such categorical programs. As such, the negative relationship suggests that either the connection between minority status and the pupil characteristics used as the basis for differential funding is not strong, or that the level of funding in this area is inadequate or sporadic across the states.

The next section examines the usefulness of using poverty level, minority status, and/or per capita income as the basis for changing the distribution of educational revenues nationwide. Such evaluation is important if pupil, district, and/or community factors are to be used as the basis for policies intended to improve the equity in the distribution of educational revenues on behalf of children across the country.

IMPROVING EQUITY

There are numerous ways to equalize spending across school districts. Table 11.5 reports the equity statistics for distributing revenues based on some of the findings from the previous analyses.

The first column of Table 11.5 lists the equity measures computed for unadjusted dollars across all the school districts in the country. On average, districts received $4,713 per pupil from state and local sources in 1989-1990. The range in average per pupil revenues was $55,900 from a low of $1,100 (in Kansas) to a high of $57,000 (in Michigan). Both of these districts are unusual: the former receives most of its funding from federal sources (Impact Aid) which is not included here, and the latter is a small district which enrolls only one student. When the restricted range is computed, the difference is reduced to $4,771 per pupil. The Federal Range Ratio indicates that spending across this restricted distribution is still 1.76 times greater at the high end than at the low. The coefficient of variation is high at 35.1 percent, and the McLoone index shows that the distribution favors those in the upper half. The Gini coefficient, which describes the degree to which equal proportions of an object are distributed over equal proportions of observations, indicates an inequitable distribution.

The remaining two columns in Table 11.5 show computed equity when the system is adjusted with reference to variables identified as important in previous analyses. For column 2, those districts where either poverty or minority enrollment is measured at or above 50 percent *and* per pupil funding is below the median, revenues are raised to the amount available to the median pupil. In other words, only those districts below the median per pupil revenues who

Table 11.5. Comparison of Equity Measures for Nationwide Unadjusted Per Pupil Revenues, 1989-1990

	1989-1990 Actual	*Adjusted for poverty and minority status*	*Adjusted for per capita income*
Mean ($)*	4,713	4,798	5,058
Median ($)*	4,451	4,451	4,451
Standard Deviation ($)*	1,562	1,519	1,298
Minimum ($)	1,100	1,100	1,100
Maximum ($)	57,000	57,000	57,000
Range ($)	55,900	55,900	55,900
Restricted Range ($)	4,771	3,982	3,717
Federal Range Ratio ($)	175.90	113.70	98.65
Coefficient of Variation (%)*	35.10	30.72	25.65
Gini Coefficient	0.172	0.157	0.139
McLoone Index	0.807	0.893	0.963

Note: *Pupil-weighted.

meet the minority and poverty criteria have their revenues adjusted upward to the median amount. For column 3, equity is computed after districts with revenues below the median and an average per capita income below $13,000 (the average in the second pupil quartile) are raised to the revenues available to the median pupil.

The results in column 2 show that compared to the actual distribution (column 1) equity improves when revenues are adjusted with reference to the minority and poverty status of children in a district. The average revenues per pupil increases $85 per pupil over the actual. The range is unaffected by the changes in the distribution, but the restricted range and the Federal Range Ratio (still high at 113.7%) improve. The coefficient of variation is 30.72 percent. The Gini coefficient moves from 0.172 (actual) to 0.157, and the McLoone index indicates a somewhat improved distribution at 0.893 (versus the actual of 0.807).

A more dramatic improvement is achieved when revenues are adjusted by per capita income. The average per pupil amount of $5,058 is $345 higher than actual. The range remains unchanged, but the restricted range at $3,717 and the Federal Range Ratio at 98.65 are improved. The coefficient of variation is reduced to 25.65 percent. The Gini coefficient at 0.139 and the McLoone index of 0.963 indicate improved distributions over the actual.

Consistent with the regression results, adjusting districts in the bottom half of the distribution based on their average per capita income makes significant improvements in equity nationwide compared with the existing distribution. Adjusting according to the percent minority and the percent of children in poverty also improves equity, but not as dramatically as when revenues are adjusted according to the average per capita income of district residents.

CONCLUSION

This study has attempted to inform discussions about improving the distribution of educational resources for all the children in the country. The focus has been on the relationship between the variation in the distribution of educational dollars and certain district characteristics.

Some surprising relationships were noted here. First, the percent of minority pupils in a district is not a completely reliable indicator of a district's revenues—high minority districts were found in the highest pupil revenue quartile. Still, the percent of minority pupils proved an important variable in explaining the variation across districts, although its influence varies by district location. Secondly, poverty is not solely an inner-city issue. Suburban and rural districts had high percentages of children living in poverty, and the vast majority of children attending schools in these districts are in districts in the lowest revenue quartile. Per capita income had the most consistent positive relationship to per pupil revenues, and was also the factor of those considered most strongly associated with the variation in per pupil revenues across the country.

When the actual revenues were adjusted according to demographic characteristics, the greatest improvement in equity was achieved using the income factor. The results of these analyses suggest that policies designed to improve equity through allocations and adjustments based on the differential characteristics of districts, their communities, and their pupils are best designed according to the capacity of the local residents to support their educational system. That is, on some measure of the income of local residents. This is, indirectly, the approach used in "power equalizing" and other equity strategies intended to remove and/or reduce the link between local fiscal capacity and revenues received. These analyses, done from a national perspective, suggest that these efforts have been less than successful. According to this research, the quality of educational opportunities for the nation's children is still a function of the ability of local residents to fund their schools. Given the differences in the ability and/or willingness of local communities and states to support their educational system, this relationship seems altogether unfair. It is also an important reality. One that it would be foolish to ignore when discussing the efficient use of educational resources to promote improved student learning.

ACKNOWLEDGMENTS

The research reported in this paper was supported by the Consortium for Policy Research in Education (CPRE) Finance Center, Grant No. OERI-R117-G40001 and by the Wisconsin Center for Education Research, School of Education, University of Wisconsin-Madison. The opinions expressed in this paper are those of the author and do not necessarily reflect the views of the U.S. Department of Education, Office of Educational Research and Improvement, the institutional partners of CPRE, or the Wisconsin Center for Education Research.

NOTES

1. The National Center for Education Statistics (1995) produced an exhaustive analysis of spending disparities across the country using the same data sources. Two important differences between the NCES study and the research presented here are found (1) in the construction of the database and (2) in the definition of the variables. The intent of the two studies is similar, but these differences make cross-study comparisons difficult.
2. New England consists of CT, ME, MA, NH, RI, and VT; Mideast consists of DE, DC, MD, NJ, NY, and PA; Southeast consists of AL, AR, FL, GA, KY, LA, MS, NC, SC, TN, VA, and WV; Great Lakes consists of IL, IN, MI, OH, and WI; Plains consists of IA, KS, MN, MO, NB, ND, and SD; Southwest consists of AZ, NM, OK, and TX; Rockies consists of CO, ID, MT, UT, and WY; and, Far West consists of AK, CA, HI, NV, OR, and WA.

REFERENCES

Berne, R. & Stiefel, L. (1984). *The measurement of equity in school finance: Conceptual methodological and empirical dimensions*. Baltimore, MD: The Johns Hopkins University Press.

Hanushek, E.A. (1996). The quest for equalized mediocrity: School finance reform without consideration of school performance. In L.O. Picus & J.L. Wattenbarger (Eds.), *Where does the money go?: Resource allocation in elementary and secondary schools* (pp. 20-43). Thousand Oaks, CA: Corwin Press.

Hertert, L., Busch, C., & Odden, A. (1994). School financing inequities among the states: The problem from a national perspective. *Journal of Education Finance, 19*, 231-255.

National Center for Education Statistics. (1995). *Disparities in public school district spending 1989-90: A multivariate, student-weighted analysis, adjusted for difference in geographic cost of living and student need.* Washington, DC: U.S. Government Printing Office, document NCES 95-300.

ABOUT THE CONTRIBUTORS

Dominic J. Brewer is an associate economist at RAND, Santa Monica, CA.

David C. Bloomfield, Executive Director of Effective Education Management, is Adjunct Assistant Professor of school law at Teachers College Columbia University and formerly General Counsel to the New York City Board of Education.

Carol E. Cohen is senior research associate with The Finance Project, Washington, DC.

Bruce S. Cooper is professor of Administration, Policy, and Urban Education at Fordham University's Graduate School of Education and a Senior Consultant at Coopers & Lybrand L.L.P., New York City.

Dan D. Goldhaber is a research analyst at CNA Corporation, Alexandria, VA.

Linda Hertert is a Director of Policy Studies at the Education Commission of the States, Denver, CO. Her areas of specialization are education policy, finance, and economics.

Hunt C. Holsomback is manager with the K-12 Education Team of Coopers & Lybrand L.L.P., Dallas office.

Warren P. Howe is on the staff at the U.S. Military Academy at West Point and a doctoral degree candidate at Fordham University Graduate School of Education.

Thomas H. Jones is professor of Education Studies at the University of Connecticut School of Education, Storrs, CT.

Larry Maloney is Director of Operations and Special Projects for the Center for Workforce Preparation of the United States Chamber of Commerce in Washington, DC.

Jay F. May is director with the K-12 Education Team at Coopers & Lybrand L.L.P., the Baltimore office.

Karen Hawley Miles is a Dallas-based consultant on school organization and finance and an advanced degree candidate at the Harvard University Graduate School of Education.

Judith S. Nappi is the principal of the Henry P. Becton Regional High School, East Rutherford, NJ, and an advanced doctoral degree candidate at the Graduate School of Education, Fordham University, New York City.

Martin E. Orland is Special Assistant to the Commissioner of the National Center for Education Statistics, Washington, DC.

Lawrence O. Picus is Associate Professor of education policy and finance at the University of Southern California, School of Education, Los Angeles.

Deborah Rosenfield is senior associate for the K-12 Education Team at Coopers & Lybrand L.L.P., Chicago.

Richard Rothstein is a Research Associate of the Economic Policy Institute, a member of the editorial board of The American Prospect, and an Adjunct Professor of Public Policy at Occidental College.

Robert M. Sampieri is Director in the Phoenix office of Coopers & Lybrand L.L.P.—K-12 Education Team.

Sheree T. Speakman is principal and partner in charge of the K-12 Education Team at Coopers & Lybrand L.L.P., Chicago.

Deborah A. Verstegen is professor of education policy and finance in the Curry School of Education, University of Virginia.

www.ingramcontent.com/pod-product-compliance
Ingram Content Group UK Ltd.
Pitfield, Milton Keynes, MK11 3LW, UK
UKHW020844290926
15470UKWH00004B/1552

9 780762 300822